Thomas Pridgin Teale

Dangers to Health

A pictorial Guide to Domestic Sanitary Defects. Fourth Edition

Thomas Pridgin Teale

Dangers to Health
A pictorial Guide to Domestic Sanitary Defects. Fourth Edition

ISBN/EAN: 9783337176723

Printed in Europe, USA, Canada, Australia, Japan

Cover: Foto ©berggeist007 / pixelio.de

More available books at **www.hansebooks.com**

DANGERS TO HEALTH:

A PICTORIAL GUIDE

TO

DOMESTIC

SANITARY DEFECTS,

BY

T. PRIDGIN TEALE, M.A.,

SURGEON TO THE GENERAL INFIRMARY AT LEEDS.

FOURTH EDITION.

NEW YORK:
D. APPLETON & Co., BOND STREET,
1883.

DEDICATED

TO

THE MEDICAL PROFESSION,

whose members, in the matter of health, public and private, have deemed it their duty—not only to restore, but to conserve,—not only to remedy, but to prevent,—not only to place their labours at the service of present suffering, but, pointing out through evil report and good report the sources of such suffering, to spend their energies in averting some of the thousand ills that flesh is heir to.

PREFACE

TO

THE THIRD EDITION.

Were any testimony needed to shew the increasing interest taken by the public in such common-place matters as drains and waste-pipes, it is to be found in the fact, that in two years and a half two editions of such a work on the subject as the present have been exhausted.

The interval which has elapsed since the publication of the first edition has given me increased opportunities of acquiring information about drainage defects. In fact, I have been a sort of centre to which such information has gravitated, and many of my friends and professional brethren tell me with pride and evident satisfaction of their sanitary discoveries, feeling that they were conferring upon me a welcome favour by telling me of some "new fault" for my book.

In this manner the material has been gathered for the increase of the number of plates from 55 to 70, and for the replacement of six of the old ones by new ones. Some additional "defects" which I could not render in picture without increasing the price of the book, some also which defied my attempts to translate into a picture, are described at the end of the book, catalogue fashion. By these additions I hope to have in some measure achieved the object which I set before me, and to have made the work a nearly exhaustive catalogue of sanitary defects, so that if ever a householder, architect, or sanitary engineer, having searched for ordinary defects, has failed to detect the fault, he may use this book as a sampler by which to test the presence or absence of extraordinary ones.

If in any case my reader or critic knows of important defects which are not here recorded, or has better advice to give, let me say with Horace :

"Si quid novisti rectius istis
"Candidus imperti, si non, his utere mecum."
Ep. I., vi., 67.

Or in the words of my friend the late Dr. Andrew Wood :

"If precepts better you should know,
"On me them candidly, I pray, bestow ;
"But if with my instructions you agree,
"At once adopt and practise them with me."

Lastly, let me express my great obligations to those who have kindly contributed information on which this enlarged edition is based, to Mr. Burton the lithographer, whose skill and care have again been at my service, and to my friend Mr. R. N. Hartley, who has given me much help and many suggestions in carrying this edition through the press.

LEEDS, *June,* 1881.

INTRODUCTION TO THE FIRST EDITION.

When, two years ago, yielding to the urgent request of the Rev. J. H. McCheane, President of the Leeds Philosophical and Literary Society, I undertook to read a lecture before that Society, and chose as the subject "Dangers to Health in our own Houses," I little thought of publishing a book, still less an illustrated book, on a subject which at first sight may appear to be outside the lines of my strictly professional work.

However, the truth of the matter is this, that having discovered and rectified one by one numerous defects of drainage in my own house, and in property under my charge, and having further traced illness amongst my patients to scandalous carelessness and gross dishonesty in drain work, I became indignantly alive to the fact that very few houses are safe to live in. Moreover, the conviction struck deeply into my mind that probably one-third, at least, of the incidental illness of the kingdom, including perhaps much of childbed illness, and some of the fatal results of surgical operations in hospitals and private houses, ("surgical calamities" Sir James Paget would call them,) are the direct result of drainage defects, and therefore *can* be and *ought* to be prevented. "Preventive medicine" has long been proclaiming such facts, and long have we turned a deaf ear, and we of the medical profession in general are only just beginning to see the great reality of her teaching.

If any one challenges this assertion in reference to my own profession, I will reply by the inquiry—How many medical men can he tell me of who understand the sanitary condition of *their own house*, or have adequately ascertained that those conditions are, as far as our knowledge at present goes, free from dangers to health? If by any possibility it could be brought about that every medical man in the kingdom should realise the necessity for looking into the state of his own house, and act upon that conviction, I feel certain that the discovery would be made in so great a proportion of instances that they were living over pent-up pestilence that we should at once have an army of sanitarians earnest and keen to ferret out unsuspected sources of illness. I take it that not a little of the lively interest recently aroused in Leeds in sanitary work may be traced to the fact that many of the medical men of this town have recently gone into the question of the sanitation of their houses, and have thereby become more keenly alive to possible sources of illness among their patients.

Hence it came about that the lecture was given which was the forerunner of this book. The lecture was delivered by request six times in Leeds, once in Knaresbro,' and once in Shipley. It was published by request of the Leeds Philosophical Society, and has had an extensive circulation.

The interest taken in the lecture and the comments and discussion to which it gave rise taught me two things:

Firstly, that if we are ever to have sound sanitary legislation, if we are ever to have our sanitary arrangements carried out in first-rate workmanship, it must be by *the education of the public in the details of domestic sanitary matters*, so that, realising their vital importance, knowing what ought to be avoided, and able to judge of the correctness and quality of work done, they may *demand* and so obtain first-rate workmanship.

When disease arises which we call "preventable," depend upon it some one ought to have prevented it.

This book will shew work defective from *ignorance*, and work defective from *dishonesty*. Probably no work done throughout the kingdom is so badly done as work in houses, drains, and pipes, which is out of sight. Probably no work is better done in the kingdom than the locomotives turned out for our railways, or the machinery which we send to all parts of the world. Are the working men less honest in the one case than in the other? I trow not. The difference is this: *Necessity* in the one case compels good work; *indifference* and *ignorance* in the other case allow bad work to pass unchallenged. If the platelayer were so to fix his rails that they would not correspond, and the next engine were thrown off the line, and death were the result, an inquest would be held, and that platelayer would be committed for manslaughter. Is there any great difference in the case where one drain pipe misses another, or ends in nothing, and in a few weeks, is the cause of death from typhoid fever? The excuse at present is that the drain layer does not know how certainly he is laying the foundation of illness and death. Disperse that ignorance, and the excuse will be gone. If the tire of the locomotive breaks, and throws a train off the line, the railway company goes to the maker of the engine, the maker of the engine to the maker of the tire, the maker of the tire to his books, and there learns the name of each foreman, and, I believe, of each workman, through whose hand the tire passed. Why can we not achieve the same connected responsibility about our drains?

Secondly, it struck me that there was need of some work of which the aim should be to teach in as simple, telling, and unmistakeable a way as possible the faults of sanitary construction which it is within the power of landlord and tenant, as distinct from the public authorities, to remedy and avoid. This latter point was pressed upon me by friends who took interest in the original lecture.

The design therefore which I have set before me is this, to represent pictorially every important fault to which domestic sanitary arrangements are liable, so far at least as my information avails me, or, in the words suggested by a medical friend, to produce "a clinical history of the defects to which drains are liable," and to point out the consequences of such defects by instances of the illness produced thereby.

In designing the illustrations one object has been kept steadily in view, viz., to give the most forcible expression I possibly could of the fact which had to be told, even at the sacrifice, if need be, of correct proportion, correct drawing, or correct perspective. This must be my general apology for the many points in which the drawings are open to unfavourable criticism.

The points in each illustration to which attention has to be attracted are drawn in strong lines, so that the eye may fix upon them first, and the lines which complete the story are drawn more faintly. The course and escape of sewer gases are indicated by blue arrows. Water in traps, water rendered impure by access of sewer gases, sewage matter in drains, and matter escaping from drains is also in blue.

If it should seem to anyone that the book is defective in that it rarely teaches how the various defects ought to be rectified, my answer is this:—

Firstly, that, when we have discovered what is wrong, we are more than half way to what is right.

Secondly, that in pointing out what is wrong, I am dealing with matters which cannot be questioned—with established and accepted principles. No one can question the fact of "a leaky joint," "a broken pipe," or "a drain running up-hill" being faulty. But in advising what ought to be done, I should be in danger of going beyond my depth, of trenching upon the province of experts, officers of health and sanitary engineers, and I should be touching on matters concerning which there may be various solutions, various opinions, and changes in course of time. What is best to-day may be superseded by what is still better to-morrow. If in any case I point out the remedy for a fault it is rather with the object, either by *contrast* to produce a more vivid impression of the original fault, or to give a *standard below which* the remedy ought not to fall. Moreover, in most instances where a remedy is suggested, a standard authority is cited for the practice.

The illustrations are planned so that each, *as a rule*, represents a single defect, and they are arranged so that the more common and obvious faults of ordinary drains come first, those which are less obvious, more rare, and more difficult to discover come next, then some of the rascalities of dishonest builders are portrayed, lastly there are added drawings as hints on ventilation, and on the exclusion of dirt from town houses and closed cases.

It is but just that I should acknowledge the kind aid without which I could not have obtained the knowledge or have produced the quality of illustrations contained in this book. My thanks are due—

Firstly, to MR. C. R. CHORLEY, Architect, of Leeds, who has superintended the sanitary alterations of my own house, has informed me of many common defects, and contributed some of the sketches from his own experience.

Secondly, to MR. ROBERT SLATER, Sanitary Inspector, of Headingley Hill, who has executed all my sanitary plumbing, has instructed me in the defects of plumbing and drains, and has informed me of defects which he has discovered in the various houses, which, owing to illness and other reasons, he has been called upon to inspect.

Thirdly, to MR. G. W. FOSTER, Artist, of Headingley, who has thrown some of my sketches into an artistic form : and lastly, to MR. WM. BURTON, Lithographer, who has executed the drawings on stone with the greatest pains and care, and has given an artistic finish to my otherwise crude sketches.

If the object aimed at has been in some degree achieved, it may be hoped that this work may be of service—

To the *householder*, who is anxious to learn whether his house is safe from drainage dangers or not, so that, aided by the diagrams, he may test every sanitary point, one by one, and, as he goes round book in hand, may catechise his plumber, his mason, or his joiner. This is the chief aim of the book.

To the *landlord*, who may learn thereby, if he does not realise them already, his responsibilities as to the health and lives of his tenants, and may feel that to save money by scamping drainage is " manslaughter under an *alias.*"

To the *medical attendant*, who may point to the pictures in the book, in order to strike conviction into the minds of his patients of the sure connection between bad drainage and ill health.

To the *architect* who may learn how by every sanitary detail which he designs amiss, or by oversight allows to be badly carried out, he is opening a door for illness to the future occupant of the house.

To the *officer of health*, who may appeal to the drawings to enforce his warnings of the dangers involved in faulty drains.

To those *entering a new house*, that they may be forewarned of the risks they run if they take the sanitary arrangements of a house on trust.

To *those about to build*, that they may know what to avoid, and what to look after, and may be able to discuss intelligently with their architect, builder, and plumber, those vital points of construction on which the health of themselves and their family will depend.

To *Town councillors* and *members of local boards of health*, that they may checkmate any of their colleagues who may have been elected to office in order to hamper or impede expenditure on sanitary work.

To *public opinion*, as one agent among many by which it is rapidly being matured, and prepared to support when the proper time arrives genuine sanitary legislation, and to demand of architects, builders, and plumbers, honest trustworthy drain work—work in matters affecting health as sound and as perfect as is now demanded and obtained in locomotives, machinery, and engineering.

Finally, let me say how fully aware I am that it is impossible in this book to include all known defects of drains, and that many omissions, probably important ones, will be discovered. Still, I trust, in a future edition, to be able to remedy any serious omissions which friends or critics may point out to me.

Leeds, November, 1878.

TABLE OF CONTENTS.

PLATE.
1. —House with every sanitary arrangement faulty.
2. —House with faulty arrangements avoided.
3. —Flame at the keyhole, and its lessons.
4. —Waste-pipe of kitchen sink untrapped.
5. —Kitchen sink trapped, discharging into gulley.
6. —Kitchen sink passing untrapped into soil-pipe.
7. —Defects in lavatories and baths.
8. —" Unsyphoned " traps.
9. —Disused traps. Evaporation.
10. — Lavatory with overflow joining waste below trap.
11. —Waste-pipe of lavatory in dressing-room passing untrapped into soil-pipe.
12. - Bedroom lavatory trapped, discharging into soil-pipe.
13. —Housemaid's sink passing untrapped into soil-pipe.
14. —Cistern feeding kitchen boiler. Overflow untrapped.
15. — Waste of bath and sink cut off, and left open to the drain.
16. —Fall-pipe carried inside house to a drain, and leaking.
17. —Fall and ventilating pipe opening near window.
18. —Vicious ventilation of drains.
19. —" Rats and the tale they tell."
20. —W.C. faulty, and with faults corrected.
21. —The " pan closet " and its substitute.
22. —" Save-all " under w.c. passing direct into soil-pipe.
23. —Soil-pipe in wall of sitting-room.
24. —Leaden soil-pipe seamed and rotten.
25. —Scullery sink discharging over dish-stone.
26. —Dish-stone admitting drain gas into larder.
27. —" Dairy sweepings."
28. —Dish-stone leading into tank under floor.
29. —Sink-pipe discharging into tank in cellar.
30. —" Sounding " for suspected tanks or cesspools
31. —Rain-water cisterns and their dangers.
32. —" How people drink sewage," No. I.
33. —" How people drink sewage," No. II.
34. —" How people drink sewage," No. III.
35. —Cesspool overflowing into a tank.
36. - Dampness of house—overflow of cesspool.

TABLE OF CONTENTS.

PLATE.
37.—New buildings over old drains.
38.—" Where *is* the butler ? "
39.—Well under house fouled by leaking soil-pipe.
40.—Square drain leaking under a tiled hall.
41.—Joints opened by settling of foundations.
42.—" Poisoned by next door neighbour's drains."
43.—" Jerry " builder buying " seconds."
44.—Drain made of " seconds."
45.—" Putty Joints."
46.—Curves made by straight pipes.
47.—Faulty junctions.
48.—Pipes laid with flange downwards.
49.—Drain running up hill.
50.—" Disconnected and misconnected."
51.—" To be continued in our next."
52.—Drain " taking " a rock; sewage " refusing."
53.—Economy in digging at expense of "fall."
54.—" Six-inch " pipe between two "four-inch" pipes.
55.—Mortar and plaster from road-scrapings.
56.—" Terrace of the future on the refuse of the past."
57.—Hunting for drains—no plans.
58.—Drain blocked by roots of trees.
59.—Cesspools under London houses.
60.—Soakings from churchyard fouling vicarage.
61.—Slop-water lodging unsuspected in cellar.
62.—Villa at a Mediterranean " health resort."
63.—An " eligible mansion " let for the summer.
64.—A Highland shooting-box.
65.—Manure heap piled against wall of house.
66.—A hint on vaccination.
67.—Poisonous wall-papers.
68.—Ventilation without dirt.
69.—Dust in cases and how to exclude it.
70.—Window ventilator in roof of brougham

PAGE.
146.—Additional Faults.
157.—Appendix, " Bye-Laws."
xvii.—Sanitary Maxims.
    Index.

SANITARY MAXIMS.

1.—It is the duty of every householder to ascertain for himself whether his own house be free or not from well known dangers to health.

2.—This duty, imperative at all times, is of surpassing urgency in a house where a woman is about to become a mother or a surgical operation is about to be performed.

3.—As a rule the soundness of the sanitary arrangements of a house is taken for granted, and never questioned until "drain-begotten" illness has broken out. In other words, we employ illness and death as our drain detectives.

4.—Whenever gas from sewers, or the emanations from a leaking drain, a cesspool, or a fouled well make their way into a house, the inmates are in imminent danger of an outbreak of typhoid fever, diphtheria, or other febrile ailment classed together under the term "zymotic," not to speak of minor illness, and depressed vitality, the connection of which with sewer gas is now fully established. Sewer gas enters a house most rapidly at night when outer doors and windows are shut, and is then perhaps most potent in contaminating the meat, the milk, and the drinking water, and in poisoning the inmates.

5.—The more complete and air-tight the public sewers of a town, the greater the danger to every house connected with such sewers, if the internal drain pipes of the house be unsound, and not *disconnected*. In houses so misconnected sewer air is "laid on" as certainly for the detriment of health as coal gas for illumination; and *you can turn off coal gas at the meter.*

6.—Every hotel throughout the kingdom, and in our watering places every house let as lodgings, ought to have its sanitary arrangements *periodically* inspected, and duly licensed.

7.—A house in which children and servants are often ailing with sore throat, headache, or diarrhœa is probably wrong in its drainage.

8.—Scamped drain-work is one of the most dangerous of the sanitary flaws of new buildings; it is also one of the most common, and one of the most difficult to detect, and is rarely found out except by means of the illness it produces.

9.—If you are about to buy or to rent a house, be it new, or be it old, take care *before you complete your bargain* to ascertain the soundness of its sanitary arrangements with no less care and anxiety than you would exercise in testing the soundness of a horse before you purchase it.

10.—If you are building a house, or if you can achieve it in an old one, let *no drain be under* any part of your house, *disconnect* all waste pipes and overflow pipes from the drains, and place the soil pipe of the w.c. *outside* the house, and ventilate it.

11.—If there is a smell of drains in your house, or a damp place in a wall near which a waste pipe or a soil pipe runs, or a damp place in the cellar or kitchen floor near a drain or a tank, let no time be lost in laying bare the pipes or drains until the cause be detected.

12.—If a rat appears through the floor of your kitchen or cellar, and a strong current of air blows from the rathole when chimneys are acting and the windows and doors of the house are shut, feel sure that something is wrong with a drain.

13.—If you are tenants, and your landlord refuses to remedy the evil, *do it at your own cost rather than allow your family to be ill.*

14.—Many a man who would be aghast at the idea of putting small quantities of arsenic into every sack of flour, and so by degrees killing himself and family, does not hesitate to allow sewer gas to poison the inmates of his house, even in the face of the strongest remonstrances of his medical adviser.

15.—A landlord may reasonably look for interest on money which he spends for the benefit of his tenant; but he is committing little short of manslaughter if, by refusing to rectify sanitary defects in his property, *he saves his own pocket at the expenses of the health and lives of his tenants.*

16.—If you be a landlord, don't intimidate your tenants or threaten to give them notice to quit if they complain of defective drainage or sewer gas in the house.

NOTE.—Copies of these "Maxims" may be obtained of the "National Health Society," Berners Street, London. Price, 2s. per 100.

PLATE I.

House with every sanitary arrangement faulty.

This plate is intended to shew at one glance the most common sanitary faults of ordinary houses. In subsequent plates each fault will, as a rule for the sake of clearness, be given singly, in order that it may be more easily understood.

A. **Water-closet** in the centre of the house.
B. **House drain** under floor of a room.
C. **Waste-pipe of lavatory**—untrapped and passing into soil-pipe of w.c., thus allowing a direct channel for sewer gas to be drawn by the fires LL into the house.
D. **Over-flow pipe of bath** untrapped and passing into soil-pipe.
E. **Waste-pipe of bath** untrapped and passing into soil-pipe.
F. **Save-all tray** below taps untrapped and passing into soil-pipe.
G. **Kitchen sink** untrapped and passing into soil-pipe.
To these might have been added a **housemaid's sink**.
H. **Water-closet cistern** with overflow into soil-pipe of w.c., thus ventilating the drain into the roof, polluting the air of the house, and polluting the water in the cistern, which also forms the water supply of the house for drinking and washing.
J. **Rain-water tank** under floor, with over-flow into drain.
K. **Fall-pipe** conducting foul air from tank fouled by drain gas, and delivering it just below a window.
M. **Drain under house** with uncemented joints leaking; also a defective junction of vertical soil-pipe with horizontal drain; the drain laid without proper fall.

PLATE I.

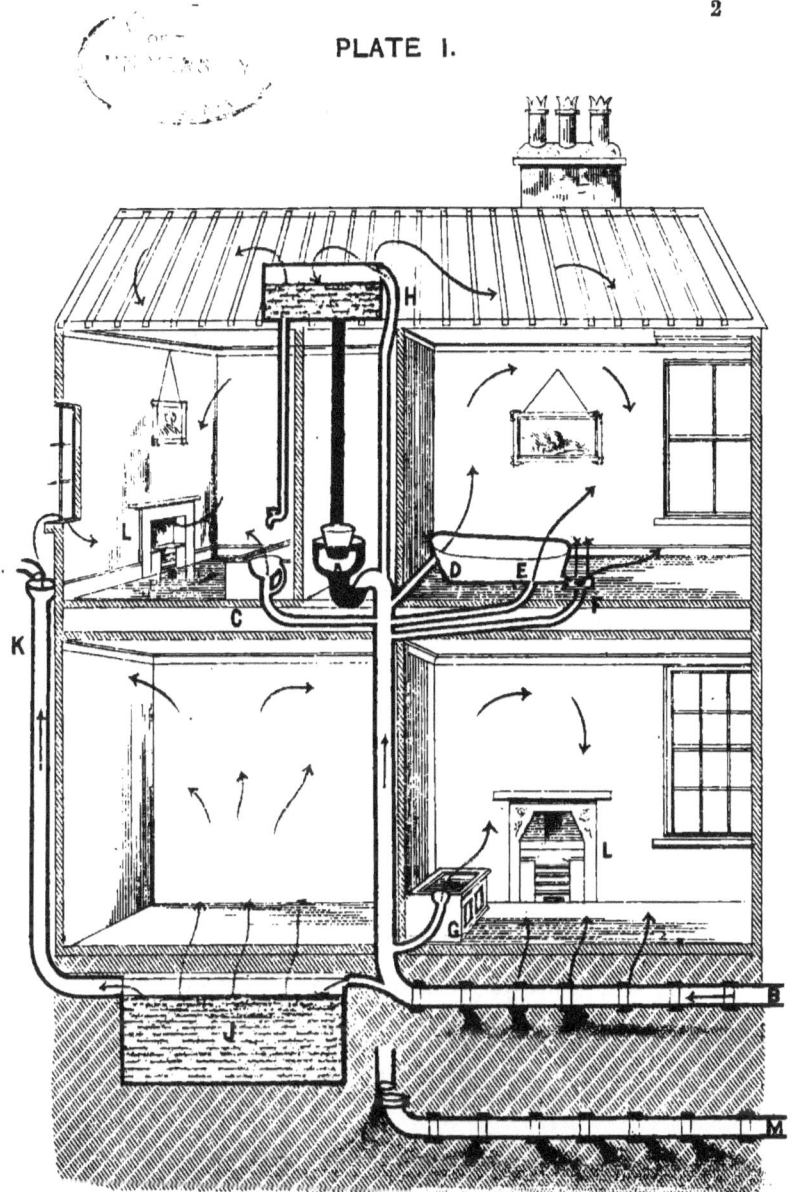

House with every sanitary arrangement faulty.

PLATE II.

House with faulty sanitary arrangements avoided.

This plate is intended to shew the reverse of the last, and to indicate the manner in which the faults can be rectified, but does not profess to lay down a strict rule as to the best arrangements.

A. **Water-closet** against outer wall of house, with soil-pipe passing directly out of the house, and ventilated by a pipe continuing the soil-pipe above the eaves, and away from chimneys or windows.

B.B. **House drains** entirely outside the house.

C. **Lavatory,**

D. **Over-flow of bath,**

E. **Waste-pipe of bath,**

F. **Save-all tray of bath,** and

G. **Kitchen sink,** to which might be added a housemaid's sink, all trapped and *disconnected* from the drain, and discharging into an open gully trap, L.

H. **Over-flow of cistern** into the open air.*

K. **Fall-pipe** near bedroom window discharging into gully L.

M. **Domestic cistern** distinct from w.c. cistern.

For more exact drawing of w.c., *vide* Plate XXI.

---

* Required by rule of Waterworks Committee of Leeds Town Council.
Building Bye-Laws of Leeds, 33f, 33i, 33j, 40, 53. *Vide* appendix.

PLATE II

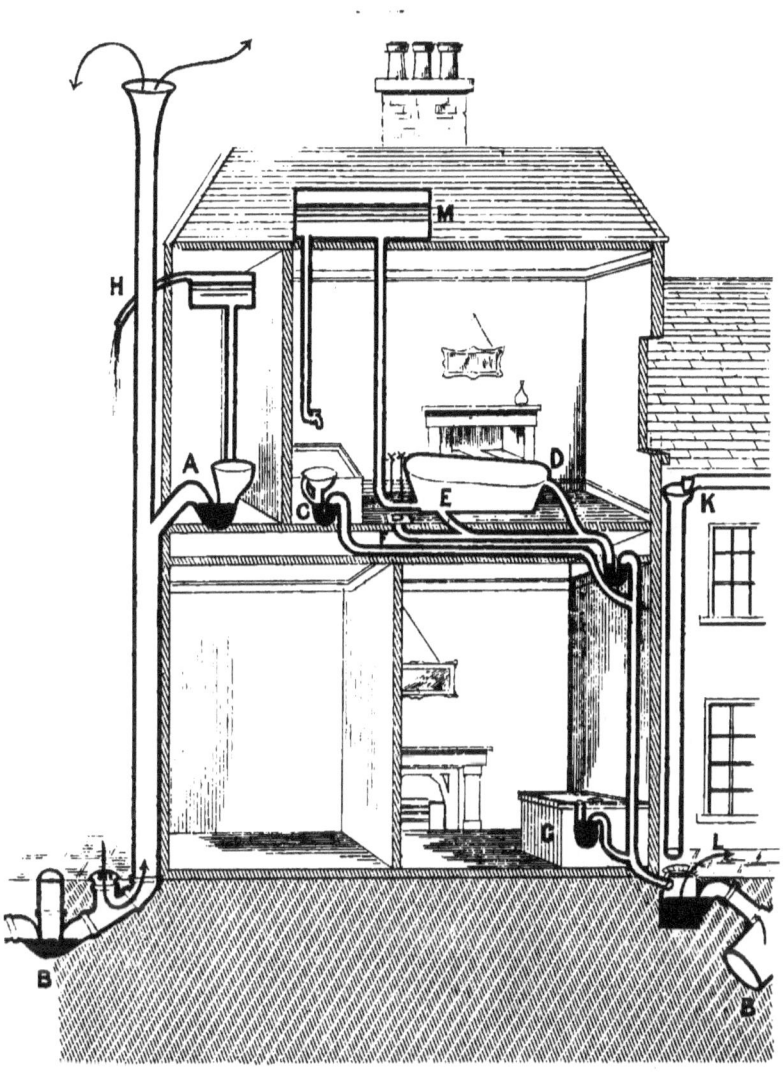

House with faulty arrangements avoided.

PLATE III.

Flame of Candle at the keyhole and the lessons it teaches.

This drawing is intended to enforce *fire* lessons :—

1st. That as a rule no provision whatever is made for the air which is to feed the chimney. An ordinary fire draws about 150 cubic feet of air per minute. If the house is well built, and the windows, doors, and floor boards fit well, the chimney smokes, unless the door or window be open.

2nd. That in the absence of any provision for the admission of air, and with the window shut, the supply of air comes from various irregular sources ; a small portion, indicated by black arrows, through window chinks ; the main portion, indicated by blue arrows, through the keyhole and crevices in the door stead, skirting boards, and floor boards. These "irregular" streams of cold air pass for the most part horizontally towards the fire, and chill the occupants of the room; and the more furnace-like the fire, the stronger the cold draught which traverses the room.

3rd. That a very moderate opening in the window is enough to stop all "irregular" draughts, the air taking the easiest course, and abandoning circuitous and contracted channels.

4th. That with a window shut, the greater part of the chimney draught is supplied from the house, and that if the air of the house be "drain-derived," then "drain-befouled" air must fill the room.

5th. That if illness "drain-begotten" breaks out in a "drain-befouled" house, and the patient cannot be removed, the safest course will be to open the *bottom* sash of the window to the extent that will allow a flame at the keyhole to burn in repose; and then to convert the horizontal draught into a vertical one by a board or cloth 6 or 8 inches high, fixed about 2 inches from the window.*

---

* Mr. F. Hinckes Bird on Costless Ventilation.—*Builder*, 1862.

PLATE III.

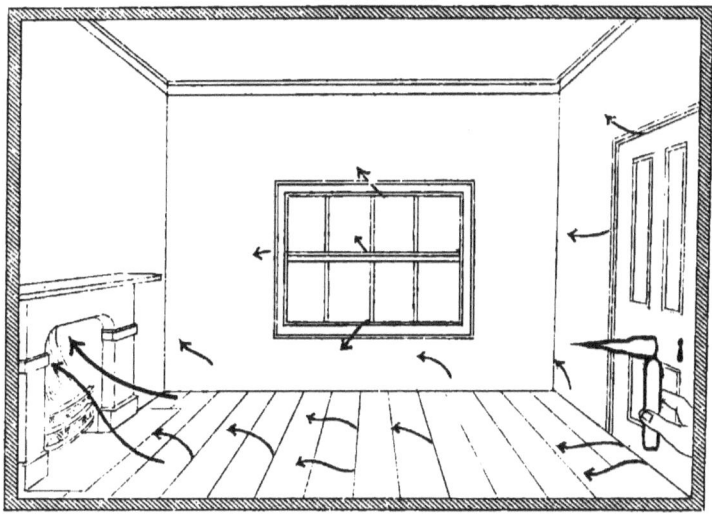

"A." Window shut. Flame at the keyhole horizontal.

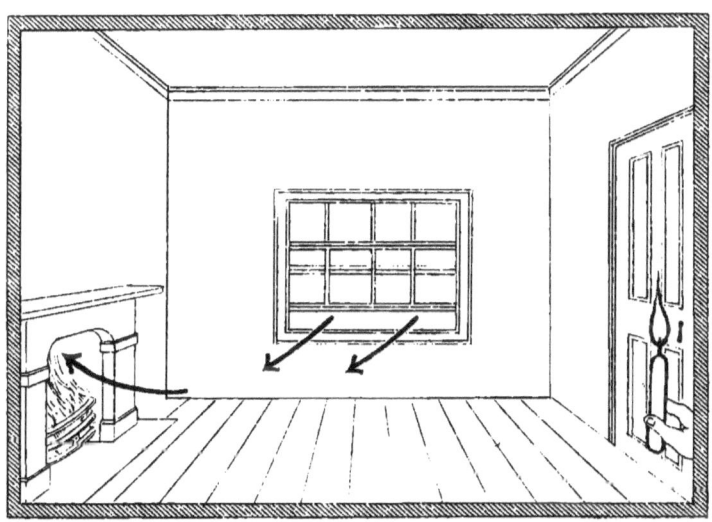

"B." Window open. Flame at the keyhole in repose.

PLATE IV.

Waste pipe of kitchen sink, untrapped, passing direct into drain.

Here are two faults—one, the absence of a syphon trap, which allows the air of the sewer to be drawn in full stream by the fires into the house, perhaps at the rate of several cubic feet per minute, and with a current strong enough to blow out a candle; the other the direct, unbroken passage of the pipe into the drain.

This is the state of the sinks of most cottages and houses which have not been recently built under the rule of "building bye-laws" of a town, or have not recently been inspected and corrected; and is almost universal in old country houses. It is probably the cause of head-ache, sore throat, and depressed health to many a cook, kitchen-maid, and butler, and perhaps indirectly leads, in not a few instances, to the use of those treacherous self-prescribed medicines—spirits and beer.

PLATE IV.

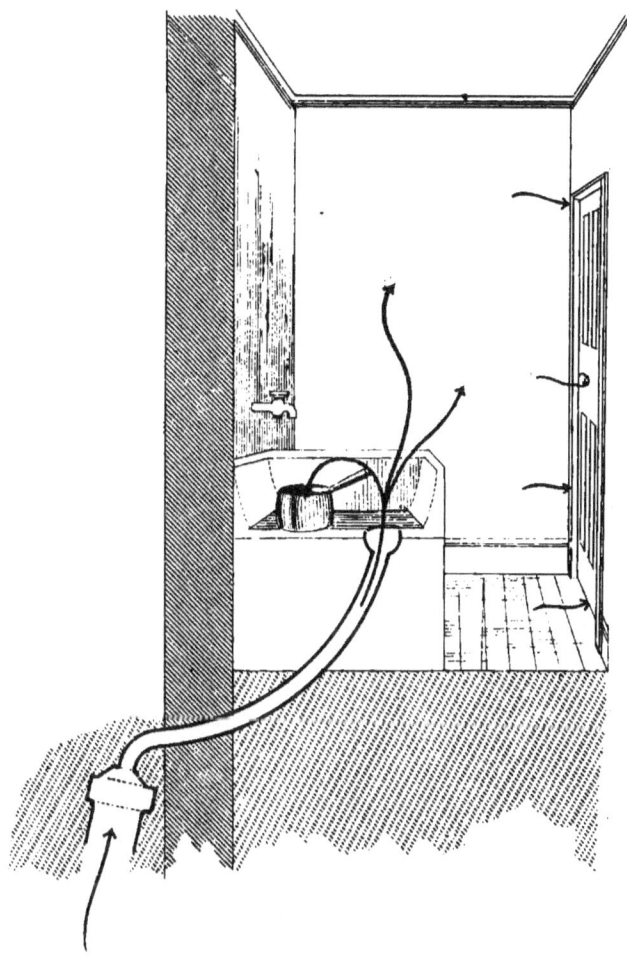

Waste Pipe of kitchen sink untrapped, passing direct into drain.

PLATE V.

Kitchen sink with faults corrected.

Fault one is corrected by a **syphon trap** A beneath the sink, which prevents any current of air being drawn into the house through the waste pipe from the surface of the water in the gulley.

Fault two is corrected by "the waste pipe being taken "through an external wall of the building into a trapped "gulley grating" (B). Building bye-laws, 33*.

What is a Gulley?

A **Gulley** is a receptacle for waste water, so contrived that, whilst it discharges its surplus water into the drain, the sewer gas is barred off from escape into the gully by the column of water filling the over-flow pipe (C).

The gulley is covered by a grating, to allow (*a*) free access of air to the surface of its contents, (*b*) the escape of any sewer gas that may be forced through the trap, and (*c*) the necessary periodical cleansing of any deposit that may collect.

What is meant by "disconnection?"

A waste pipe is said to be "**disconnected**" when, instead of being continuous with the drain pipes, it discharges into a gully, *i.e.*, practically into the open air.

In the drawing the waste-pipe delivers into the "gully" below the grating, as a precaution against frost. Some authorities insist upon the pipe delivering above the grating.

What is here said of the waste-pipe of a sink equally applies to the waste-pipes of baths and lavatories, and housemaids' sinks. For disconnection of w.c. and soil pipe, see Plate XXI.

PLATE V

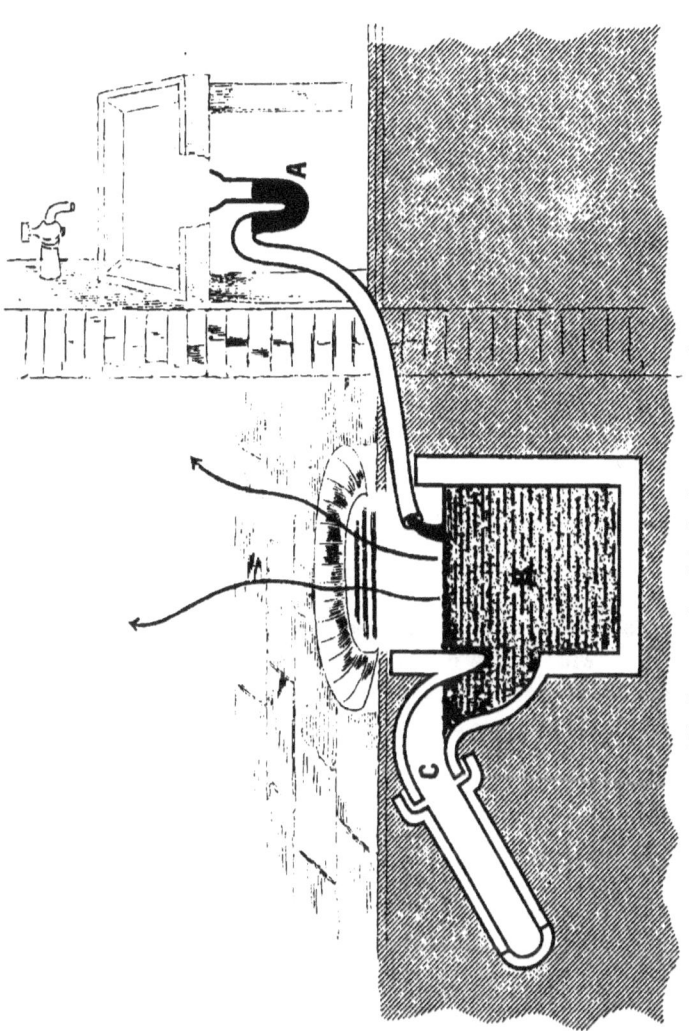

Kitchen sink with faults corrected.
How to "disconnect" sinks, baths, and lavatories.

PLATE VI.

Waste pipe of kitchen sink, untrapped, and passing into soil pipe.

This was found in a house recently occupied by a relative of my own. The w.c. soil pipe being conveniently near, had been tapped by the ignorant or indolent plumber to receive the waste pipe.

PLATE VI.

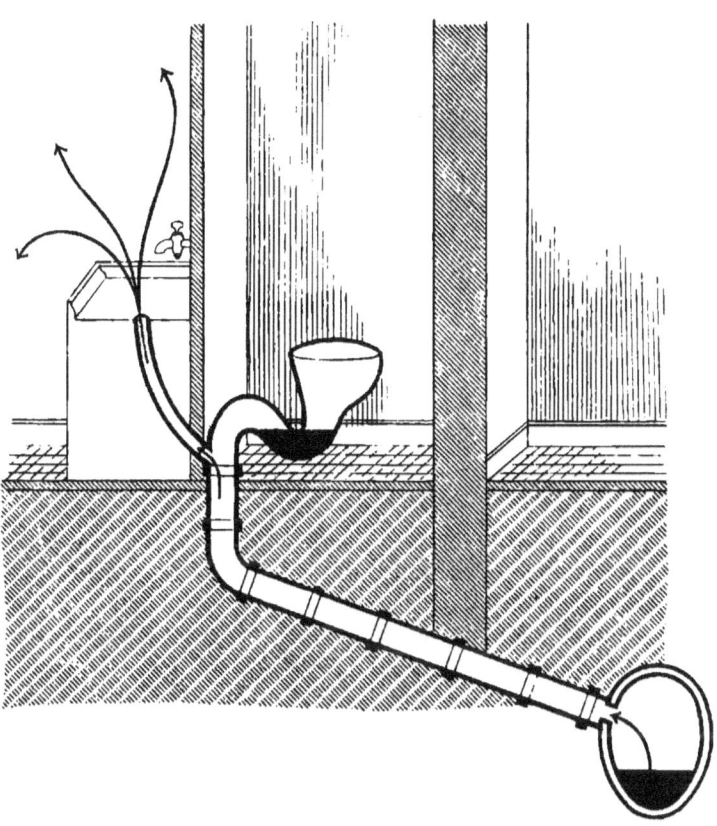

Kitchen sink carried untrapped into soil pipe of w.c.

PLATE VII.

Defects in lavatories and baths, and their remedies.

A. Waste-pipe of lavatory, waste and over-flow pipe of bath, all untrapped and passing into soil-pipe of w.c.
B. Lavatory waste-pipe trapped and discharging into open gully outside the house. (Building bye-laws). Waste-pipe of bath also remedied, but the "over-flow" is still untrapped and joins the soil-pipe. It is not uncommon to find that, the waste-pipe being trapped and delivering into a drain or gulley, after a while the bath over-flows. Another plumber is then called to add an over-flow pipe, who, ignorant of his business, takes the over-flow untrapped into the nearest communication with a drain, which is usually the soil-pipe of a w.c.
C. In this drawing, both waste and over-flow of bath are properly guarded by a trap, and properly conducted into the open air, but by an oversight the "save-all" tray for catching the drippings of the taps has been connected directly with the soil-pipe, thus vitiating the whole arrangement. This fault was recently discovered close to the bedroom of a gentleman suffering from whitlow with inflammation spreading up the arm, his medical man having insisted on a close investigation of the drains of the house under the conviction that some such cause was needed to explain the attack.
D. All pipes from bath correctly arranged.

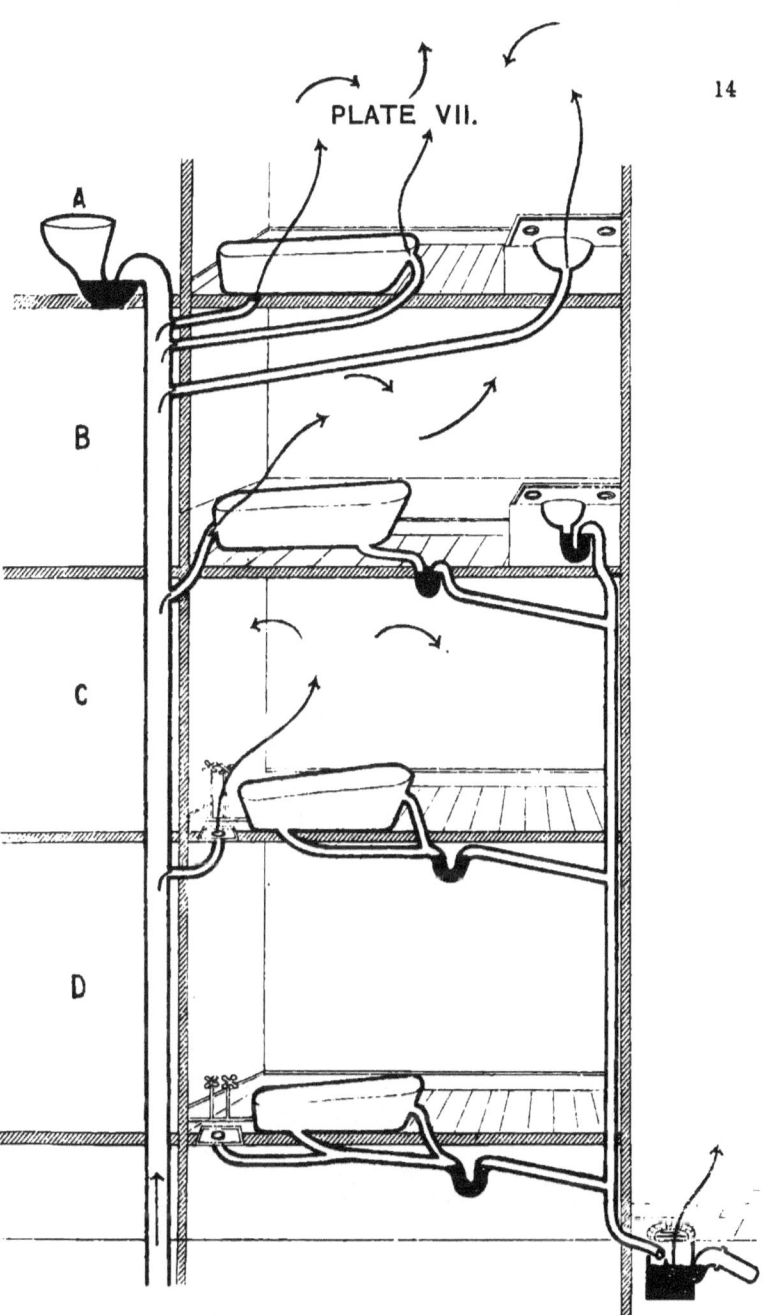

PLATE VII.

Defects in lavatories and baths, and their remedies.

PLATE VIII.

"Unsyphoned" Traps.

This is an attempt to suggest in a diagram the effect of water in motion.

When the water is being run off from the bath (B,) the falling column of water as it rushes past the entrance of the pipe of the lavatory (C) sucks the water out of the trap of the lavatory, "unsyphons" it, and leaves it open to the drain until more water is let in to fill the trap.

The same is said to occur in the case of water-closets, (FDE) where a series, one above the other, discharge into the same soil-pipe, an arrangement more common in London than elsewhere.

What is the remedy?

Let me quote from Mr. J. A. Russell's lectures to Plumbers and Builders, page 19.*

"6th. Traps may be unsyphoned by a body of water "coming down the soil-pipe from a fitting higher up on the "same stack. Such a body of water will act like a piston, "compressing the air in front of it, and making suction "behind it. One gallon of water fills nearly $39\frac{1}{4}$ inches of "3 inch pipe, 28·8 of $3\frac{1}{2}$ inch, 22 of 4 inch, and 17·43 of "$4\frac{1}{2}$ inch. The remedy is to have a ventilating inlet joining "the highest point of the bend on the distal side of the trap, "and if the vent be taken from the soil-pipe higher up, (and "not from a separate air pipe, or a grating to the open air,) "the above data will indicate the proper distance."

---

* Sanitary houses, by J. A. Russell, Lecturer on Sanitation at the Watt Institution, Edinbro'.—*Macluchan and Stewart.*

PLATE VIII.

"Unsyphoned traps."

PLATE IX

Disused Traps; Evaporation.

Traps cease to be traps as soon as the water evaporates below "the seal."

Unoccupied houses are liable to have open communications with the drains from this cause; and lavatories, and water-closets rarely used may become "unsealed" from disuse and evaporation.

It is not uncommon to hear people say, "oh, we never use "such and such a w.c. except in case of illness," forgetting that disuse means evaporation, and open communication with a drain. Probably much illness has resulted from evaporation of the water in the syphon of a lavatory of a seldom used "spare bedroom."

PLATE IX.

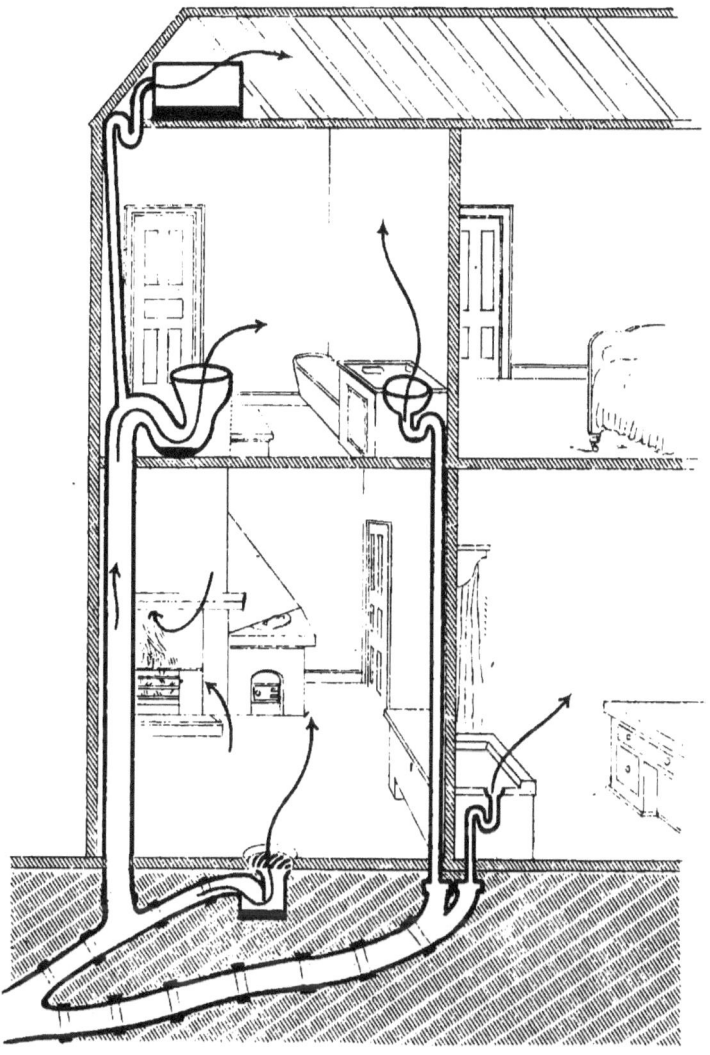

Disused Traps.—Evaporation of water and consequent direct communication with the drain.

PLATE X.

Lavatory with overflow joining the waste-pipe below the trap.

In this instance (not very uncommon, though a violation of common sense) the trap was rendered useless. because the over-flow (A), missing the trap (C), communicated directly with the drain (B), and served as a ready channel for the passage of sewer gas.

It was discovered in the house of Mr. E. Atkinson, surgeon, of this town, a house sold to him as recently fitted up with all precaution as to sanitary requirements.

PLATE X.

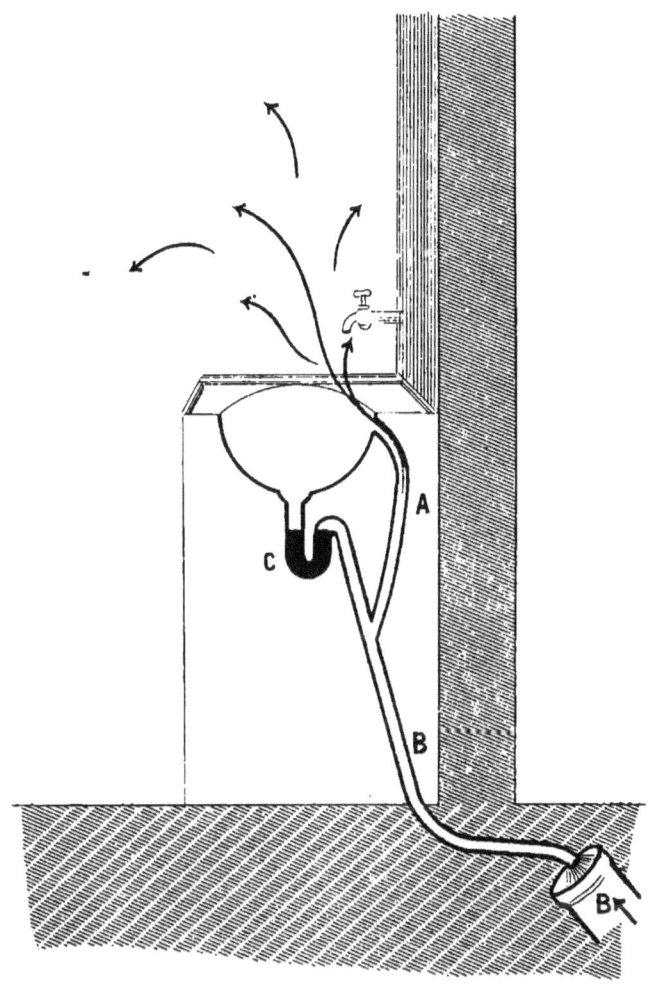

Lavatory with overflow joining waste pipe below the trap.

PLATE XI.

Waste-pipe of lavatory in a dressing-room passing untrapped into a drain or soil-pipe.

This condition along with other faults was discovered in the house of a medical man whose wife had been dangerously ill from puerperal fever. Her accouchement being again in prospect the husband very wisely had the sanitary condition of the house enquired into, and this and several other serious defects were discovered. All was set right, and on this occasion the lady recovered without a drawback.

PLATE XI.

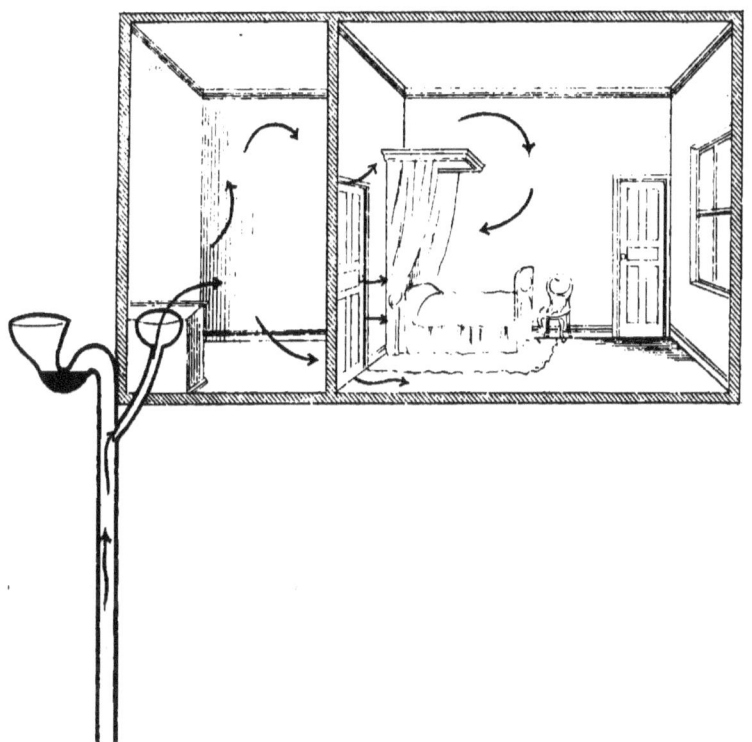

Lavatory in a dressing-room opening out of a bedroom with waste-pipe untrapped and connected with soil-pipe.

PLATE XII.

Lavatory in bedroom trapped, but discharging into soil-pipe of w.c.

The syphon trap (A) prevents any rush of air being drawn through the waste-pipe (B,) but does not prevent the slow passage of foul gases from the w.c. drain, (C,) indicated by the faint arrows rising from the basin. The gentleman occupying the bedroom from which this illustration was taken, was suffering from erysipelas of the face, and was about to undergo a surgical operation. His surgeon refused to do any operation until the lavatory pipe was cut off from the drain, and made to discharge into the open air. It is right to add that the w.c. was in the centre of the house, and that the drain ran under the hall floor.

PLATE XII.

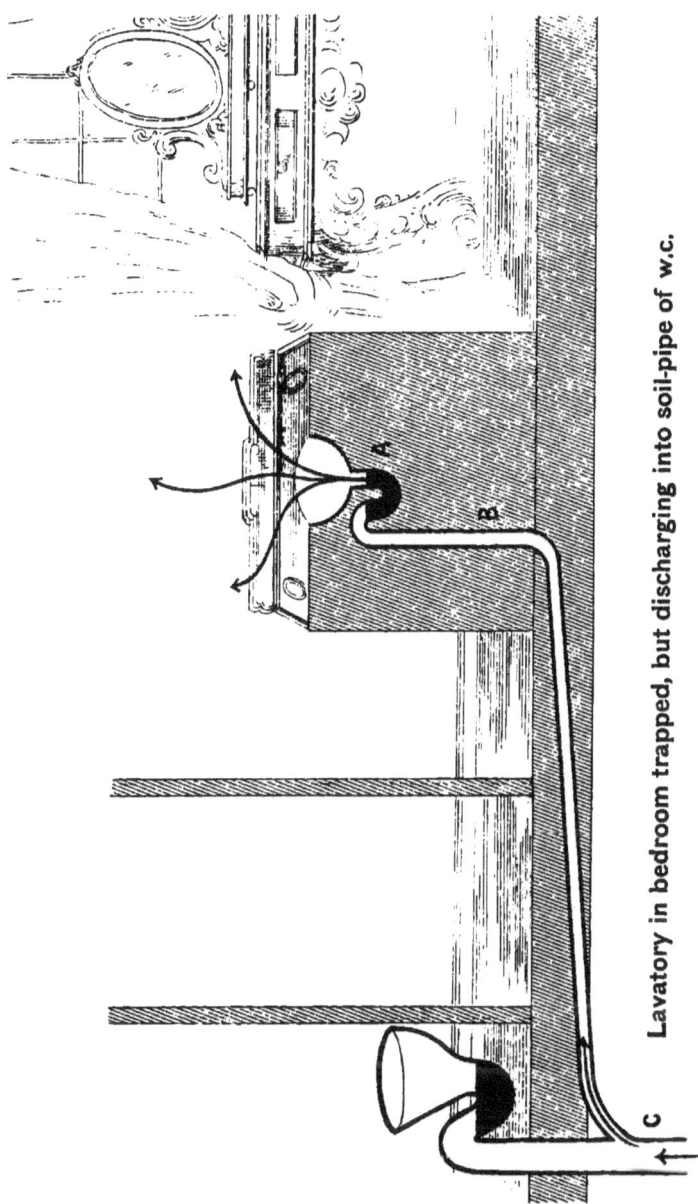

Lavatory in bedroom trapped, but discharging into soil-pipe of w.c.

PLATE XIII.

Housemaid's sink-pipe untrapped and discharging into a soil-pipe.

This plate seems but a repetition of the untrapped lavatory, but is introduced because the housemaid's sink, often in a dark corner, is apt to be overlooked even when all proper care has been taken with lavatories and baths.

This instance is communicated to me by Mr. Nicholson Price, surgeon, of Leeds. He had recently removed to a house the property of the Leeds Infirmary. In three or four months two of his children became seriously ill with inflamed throat. The sanitary condition of the house was suspected and investigated, and it was found that two housemaids' sinks near the bedrooms passed virtually untrapped into soil-pipe and drain.

PLATE XIII.

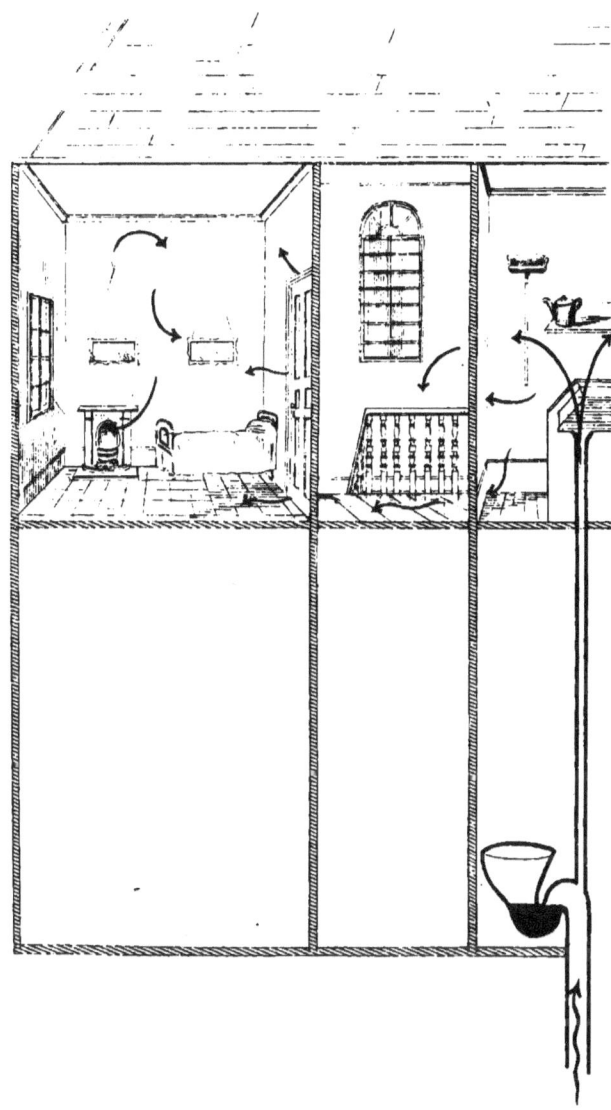

Housemaid's sink pipe passing untrapped into soil pipe.

PLATE XIV.

Cistern feeding kitchen boiler.

Where a cistern is arranged to feed a kitchen boiler, the cistern must have an overflow pipe. This overflow pipe is often carried direct into a drain without even the partial protection of a trap, and thus establishes a channel for sewer air to come into the kitchen, and to foul the water of the boiler. If water for the kettle be drawn from the boiler, then impure water is drunk.

PLATE XIV.

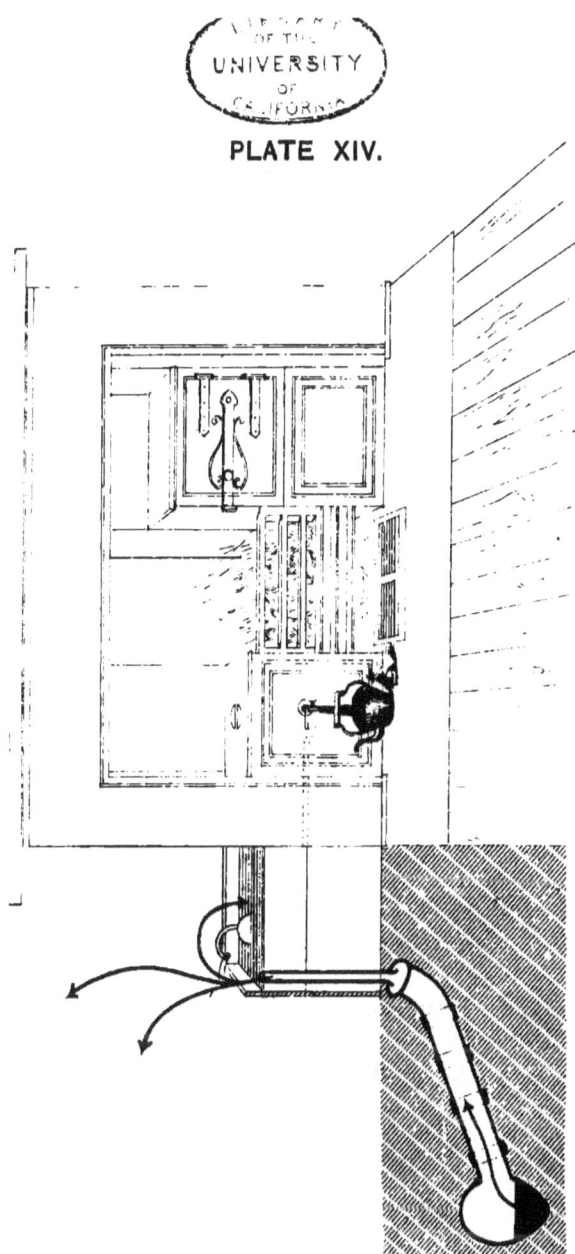

Cistern feeding kitchen boiler with overflow direct into drain.

PLATE XV.

Waste-pipe of bath and sink cut off —pipe open.

(A) was discovered in the following manner :—
Mrs. A., from Lancashire, came to spend a few days in Leeds. Soon after her arrival she consulted a medical man about a severe neuralgia of the face and side, and complained of a sore throat. The medical man on seeing her throat, at once enquired about her drains, but could not discover that anything was wrong. In three days she reported herself as cured by the remedies prescribed. Her doctor thinking the cure too rapid to be the result of his medicine, again catechised her about her drains, and at last drew out that there had been a bath in a room near her bedroom, that the bath had been removed, but that the waste-pipe had been *left* (open of course) *in case they might wish to replace the bath.* The room had been constantly so unpleasant, that an apprentice who slept there had his window open summer and winter, and they had made many fruitless efforts to discover the cause.

(B) is taken from No. 20, Park Row, the house I formerly occupied. The scullery, on my leaving the house, was turned into an office, and the sink was removed. A few years after, the clerks complained of bad smells, and after much search the cut off waste-pipe of the sink was discovered underneath the floor boards open-mouthed, and passing direct into a drain.

PLATE XV.

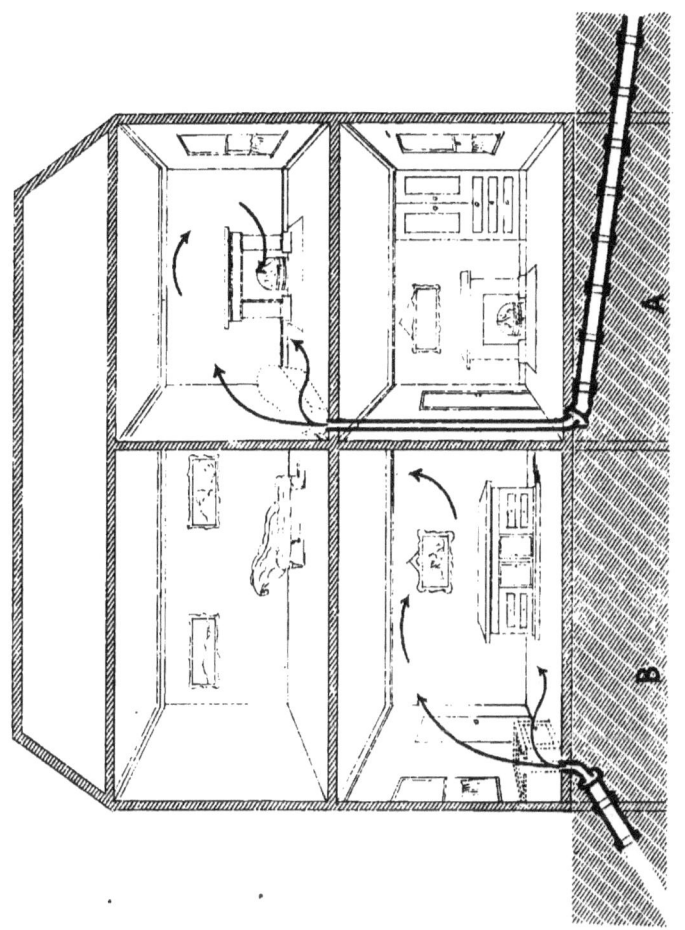

A. Disused bath, waste pipe cut off and left as an open communication with the drain.
B. Sink waste pipe treated in a like manner.

PLATE XVI.

Fall pipe having direct communication with the drain carried through the house and allowing the escape of sewer gas from imperfect joints.

This house, at a watering place, was first tenanted and afterwards purchased by a relative of my own, who after a residence of a few weeks, had erysipelas of the face. This attack at once suggested to me drainage faults, and made me reproach myself for not having had the house previously inspected. An inspection discovered the rain-fall pipe carried from the front through the cellars into a drain at the back. The joints of the pipe as it passed through the house were so defective that pans had to be placed to catch the rain. A bath upstairs had a waste pipe opening untrapped into the fall pipe. There was also an untrapped sink in the kitchen. After purchase of the house, the defects were remedied, and all pipes were *disconnected* from the sewer.

Mr. C. R. Chorley tells me that in one of the Yorkshire country mansions which he inspected, he found, along with numerous other faults, that *all the fall pipes had been carried, for the sake of appearance, inside the walls*, actually in the corners of bedrooms direct into the drains, and that the joints inside the house were incompetent and open, and allowed the plentiful escape of sewer gas.

Recently in my own consulting rooms, built for me six years ago before I gave much thought to drains, Mr. Chorley discovered a rain-fall pipe carried through the centre of the house into a drain. The existence of this I had not suspected. The defect was remedied by conveying the pipe to a gully on the outside of the house.

PLATE XVI.

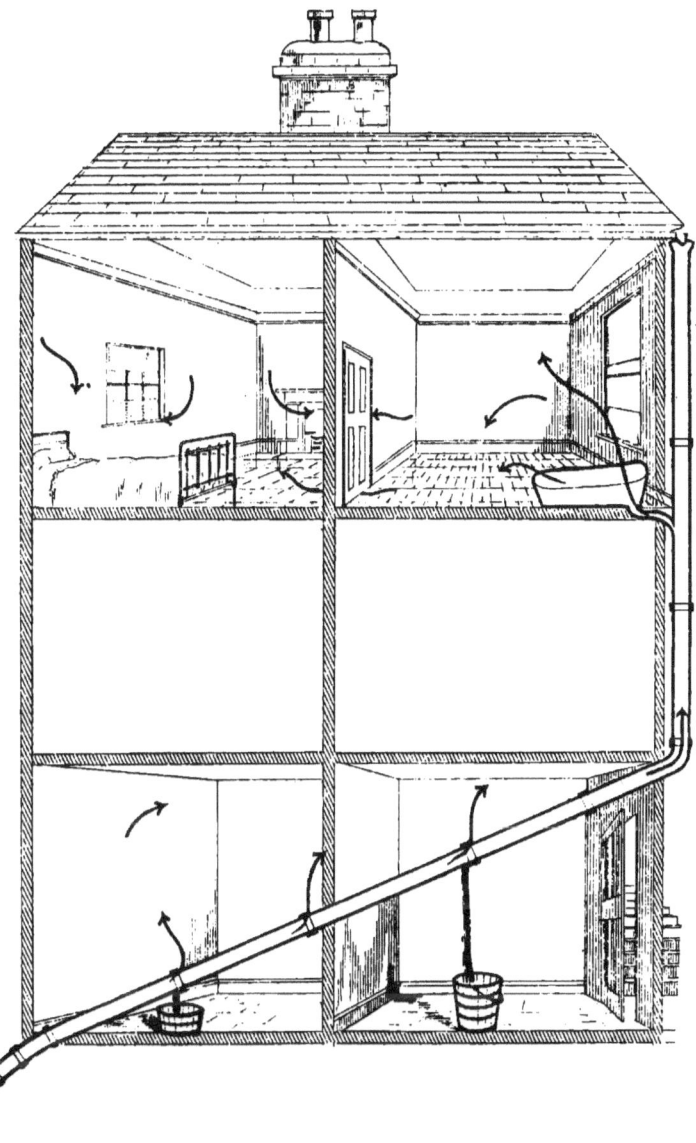

Rain fall pipe with leaky joints carried through basement direct into a drain.

PLATE XVII.

(A) Fall pipe communicating with sewer and opening just below bedroom window.

(B) Ventilator of soil-pipe opening below attic window.

This is a not uncommon, though often an unsuspected source of danger. Some years ago, an outbreak of typhoid fever in one of the colleges at Cambridge was attributed to this cause.

Of the same class of faults are those arrangements common in London houses, in which a leaden roof over an outbuilding or bay-window, or a cistern outside a window, have fall pipes, or overflow pipes passing into drains.

This ventilating shaft is faulty in two points—(1) in not being as large as the soil-pipe, (2) in its termination, a trumpet-shaped opening, as in Plate I., being deemed the best.

PLATE XVII.

A. Fall pipe communicating with sewer, and opening just below bedroom window.
B. Ventilator of soil pipe discharging close to attic window

PLATE XVIII.

Vicious ventilation of drains.

A. A fatal case of typhoid fever occurred at a school in which sanitary precautions had been taken with great care and anxiety by the schoolmistress. This led to a fresh investigation, and to the discovery that the studiously planned ventilation of the drain had been ingeniously mismanaged, and that the "ventilator" had been turned into the chimney of the room in which the young lady slept.

B. The ventilating shaft from a drain ought not to end near the top of a chimney, lest the sewer-gas be carried by a down-draught through the chimney into the house.

PLATE XVIII.

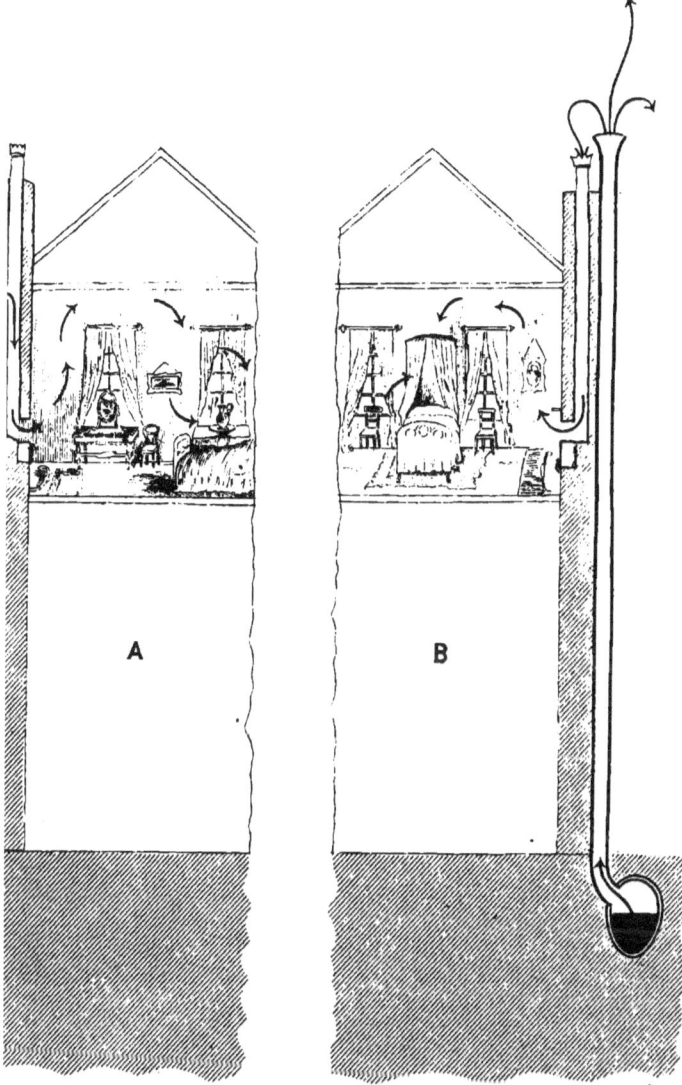

A. Ventilating pipe of drain turned into bedroom chimney.

B. Ventilator of drain discharging close to a chimney-pot.

PLATE XIX.
"Rats, and the tale they tell."

When rats appear in a kitchen or cellar the presumption is that they come out of a drain. A hole in a drain which permits the escape of a rat will allow the sewer gas to be drawn into a house : *"pleno flumine."*

When a waste-pipe or a sink joins a drain under a kitchen floor instead of discharging into a gulley outside, this is what usually happens. The sink-pipe *religiously trapped* passes *neatly through the kitchen floor.* Beneath the floor and out of sight it passes into an open wide-mouthed drain pipe, 4 or 6 inches in diameter, with neither cement nor luting to bar the escape of rats or sewer gas. This piece of scamping being out of sight is *exceedingly common*, and is often overlooked by Inspectors who satisfy themselves with a peep at the syphon trap, and take no account of the gaping pipe concealed beneath the flag, ready to let the rat and the gas out of the drain.

This was discovered in a house which I recently bought.

In my own kitchen also a flaw of this kind was found. The cement forming the junction of the sink-pipe and drain was eaten or broken away, leaving a hole large enough to receive a man's hand.

I need hardly say that I had the sink-pipe turned into an outside gulley, and the drain under the kitchen entirely removed.

In two other ways rats do mischief—one, by eating through lead pipes in order to reach water or fat—the other by making runs under drain pipes and letting down and opening the joints.

Open drain joints concealed under a cellar-floor can often be detected in the following way :—shut all windows and outer doors—open all doors between the cellar and the fires in the house—then hold a lighted taper opposite any crevices or fissures, such as are shewn by the blue arrows.

PLATE XIX.

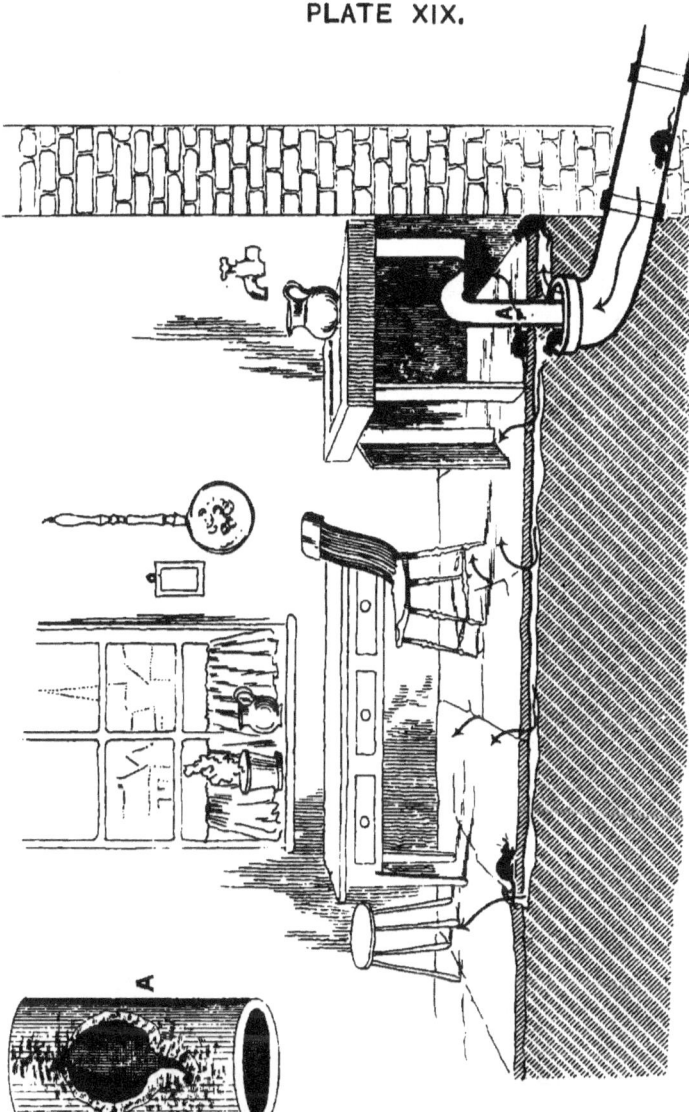

Rats and the tale they tell. AA. Hole in leaden pipe gnawed by Rats.

PLATE XX.

Water-closet with arrangements all faulty, compared with w.c. with the faults remedied.

As the arrangement of the w.c. is to many persons a source of great anxiety, I have felt obliged to depart somewhat from the rule laid down, and to suggest a plan which seems to be free from serious objection, and which has been adopted in my own house.

In No. 1, the pan (A) is a "pan-closet" very common, and objectionable because of the large cavity—or so called "container"—between the pan and the trap. This cavity becomes foul, and a receptacle for foul air, which either passes through the water by absorption, or is displaced into the house when the closet is used. In No. 2, the "pan-closet" is replaced by a simple syphon sanitary basin (C).

In No. 1. the soil-pipe (B) is inside the house, and if faulty at any part, allows the escape of dangerous gas into the house.

The soil-pipe may be faulty,—

(*a*) At the junctions with the pan above, or the drain below, from the joints being badly made, "putty joints" instead of soldered joints, or the pipe may have settled, and so have opened the joints;—

Or, (*b*) The lead pipe may be "seamed" instead of "drawn," and so liable to gape at the seam;—

Or, (*c*) The lead pipe may be old, twenty or thirty years, and eaten through by the sewer gas;—

Or, (*d*) The soil-pipe may be made of short sanitary tubes, affording many joints for insecurity and the escape of sewer gas;—

Or, (*e*) The soil-pipe may by its weight have broken the earthenware junction with the drain, thus allowing the discharge of the sewage beneath the floor of the house. *Vide* Plates I., **XXXIX., XLIX.**

PLATE XX.

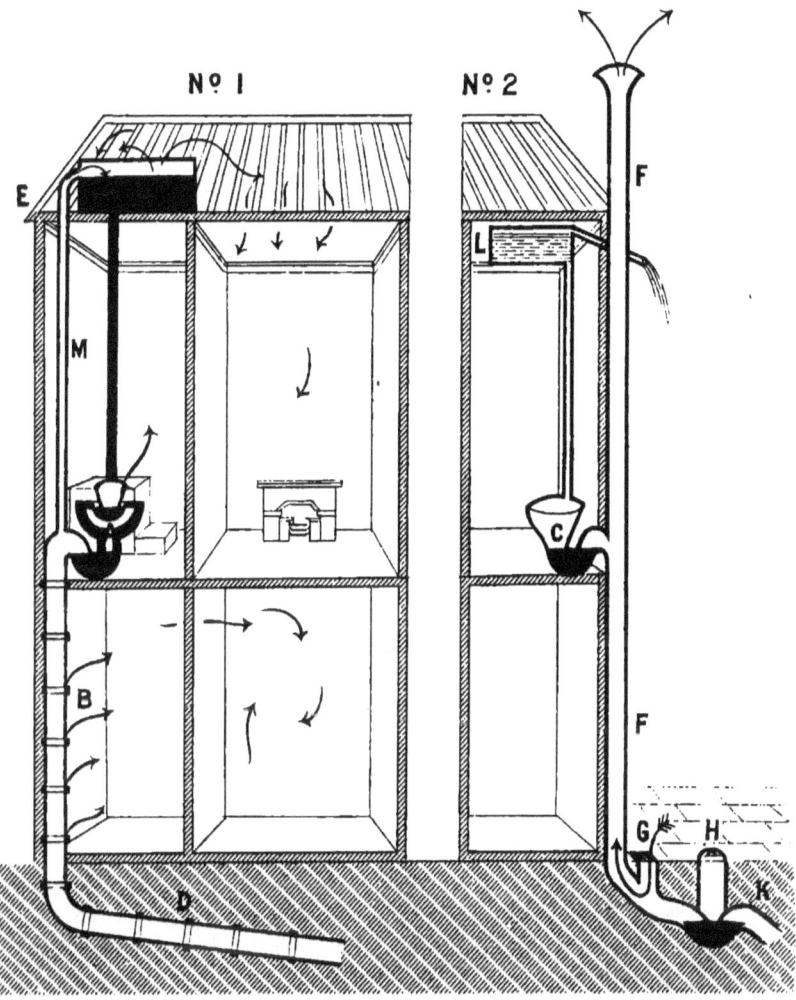

No. 1, w.c. faulty.   No. 2, faults corrected.

PLATE XX—(Continued).

In No. 2, these risks are avoided by carrying the soil-pipe (F) outside the house, to join an outside drain.

In No. 1, (D,) the drain is underneath the house, and if it is laid wrongly, without proper fall, (Plate I., XLIX.,) or badly, with unluted joints, (Plate I.,) or of broken, *i.e.* "seconds" pipes, (Plate XLIV.,) or if the foundation sinks, (Plate XLI.,) a cesspool is formed at every leaky point within the house, (Plate I.)

In No. 2, the drain (K) is entirely outside. On "drains under any building," compare Building Bye-laws, § 33f.

In No. 1, the cistern (E) has its overflow into the soil-pipe, thus acting as a ventilator to the drains, and conducting the sewer gas into the roof, and thence into the rest of the house.

In No. 2, the overflow of the cistern (L) discharges into the open air—in accordance with the bye-law of the Waterworks Committee, of the Leeds Town Council.

In No. 1, the soil-pipe is unventilated except by the overflow pipe of the cistern (M).

In No. 2, the soil-pipe (F) is "continued upwards without diminution of diameter," above the eaves "to such a height and in such a position as to afford, by means of the open end of such pipe, a safe outlet for sewer air," or in other words the ventilating pipe must not end anywhere near a window, (see Plate XVII.,) nor a chimney top, (Building Bye-laws, § 53,) (Plate XVIII.)

In No. 2, there is an open air grate (G,) to allow the free passage of air up and down the soil-pipe, and to prevent the accumulation of foul gas on the drain side of the water trap of the w.c. basin.

In No. 2, there is a syphon trap (H) to cut off the sewer gas from the soil-pipe, with a tube closed by a moveable top by which access can be gained to any stoppage in the trap.

PLATE XX—(*Continued*).

Besides all this, there must be a ventilating tube on the drain side of the syphon trap (H).

On the subject of "ventilation of drains," Dr. Clifford Allbutt tells me of a case of typhoid fever attended by himself and Dr. Dobie, of Keighley, "due to the magnificent completeness of the whole drainage, done at great cost, including an equally magnificent cesspool, 300 yards away, and all absolutely tight, and so *unventilated* anywhere"— except into the house through the water-traps.

PLATE XXI.
The "pan-closet" and its substitute.

This Plate is introduced at the risk of repetition for two reasons,—

Firstly, in the hope of giving the *coup de grace* to that utter abomination of sanitary mechanism, condemned by all sanitary reformers, the "pan-closet." If by giving it a nickname one can aid in rendering it impossible for architect or builder to insert one in a new building, without incurring ridicule,—*Ridiculum acri Fortius et melius magnas plerumque secat res*—an additional blow will have been struck at a source of offensiveness, if not of illness, but too common even in well appointed houses.

This contrivance is vividly described, and no less scathingly condemned, in an article on "The Sanitary condition of New York," in "Scribner's Monthly Magazine" for May, 1881, page 74, from which the following is an extract:—

"The cardinal fault of all, not even surpassed by the "unventilated soil-pipe, is the w.c., which is in almost universal "use all over christendom. This is known as the "pan-closet." "It probably is not, but it certainly might be, the invention of the "devil."

Secondly, because the former editions of this work have been taken much more as a guide to sanitary arrangements in the construction and re-construction of dwellings than I contemplated, and in consequence I feel bound to give a somewhat detailed drawing of what seems to be the simplest plan of w.c. which would be deemed satisfactory by sanitary engineers.

The **closet** should consist of a simple **basin (A)** and **syphon (B.)**, the syphon being cleansed each time of use by a **strong two gallon flush.**

A receptacle for water in the basin in addition to the syphon is a complication, and may **reduce** the force of the flush upon the syphon unless skilfully managed. The cleaning of the syphon by the flush is the main point to be solved.

PLATE XXI.

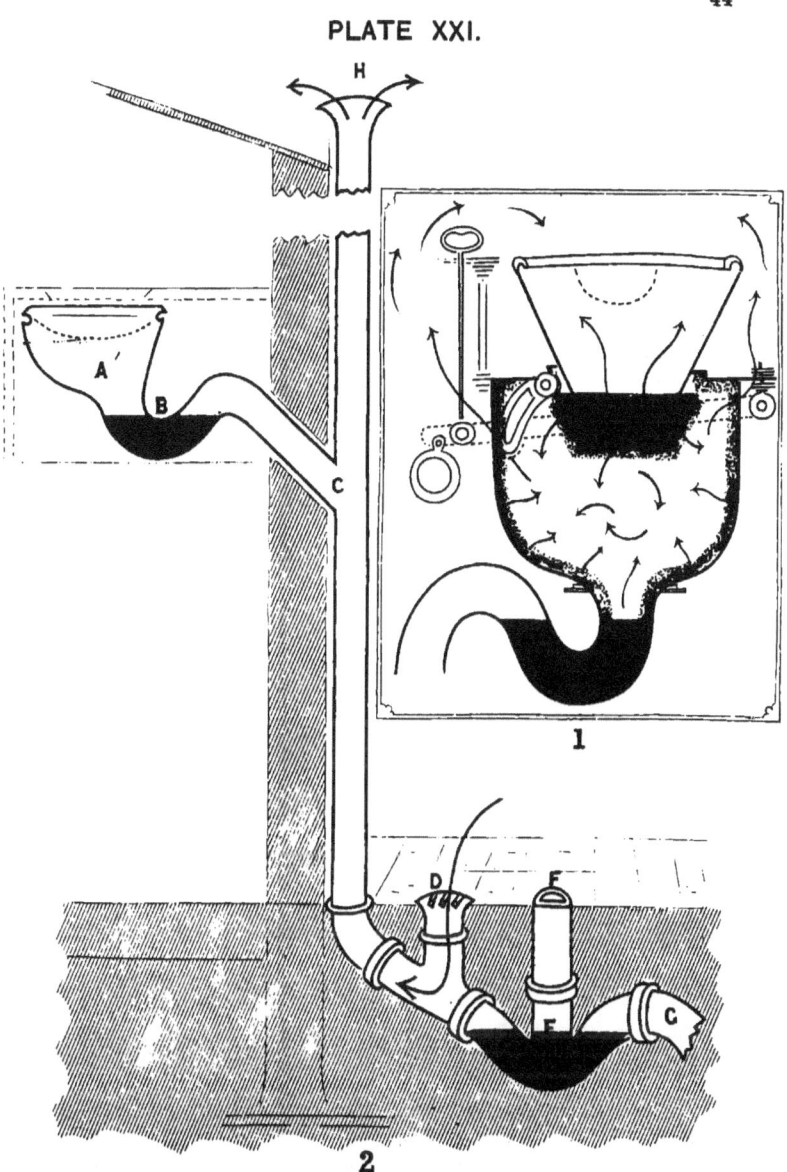

1. The "**UTTERLY UTTER**" abomination.
2. The modern substitute.

PLATE XXI—(Continued).

The syphon should not be of greater calibre than the lower orifice of the basin, and should have a very shallow water seal, and a good, not an abrupt, curve, in order that the force of the water may be as effectual as possible in clearing it out.

The soil-pipe (C) should be outside the house, open at the top (H) above the eaves, and open at the bottom (D) at or near the ground level (as represented by a black arrow), and, finally, cut off from the sewer by a syphon trap (E).

The syphon trap should communicate with the surface of the ground by a tube (F), sealed at the top by a moveable plate for the purpose of cleansing in case of need.

Finally, let me say that I have had these arrangements in use for several years, with perfect satisfaction and no inconvenience from frost.

PLATE XXII.

"Save-all" tray beneath w.c. with untrapped waste pipe serving as unsuspected ventilator to soil-pipe.

This drawing was communicated to me by Mr. C. R. Chorley, who discovered it in a house in which very great pains had been taken with the waste pipes and drains. The "save-all" (A) is sometimes placed under a w.c. to catch any chance overflow when slops are carelessly emptied into the pan, and the waste pipe (B) is, naturally perhaps, but most disastrously, carried untrapped into the soil pipe (C). Even a trap to the waste pipe of the tray is a "snare," as Dr. Clifford Allbutt said, because it only acts when there has been great carelessness resulting in an over-flow sufficient to fill the trap, and this will soon evaporate.

PLATE XXII.

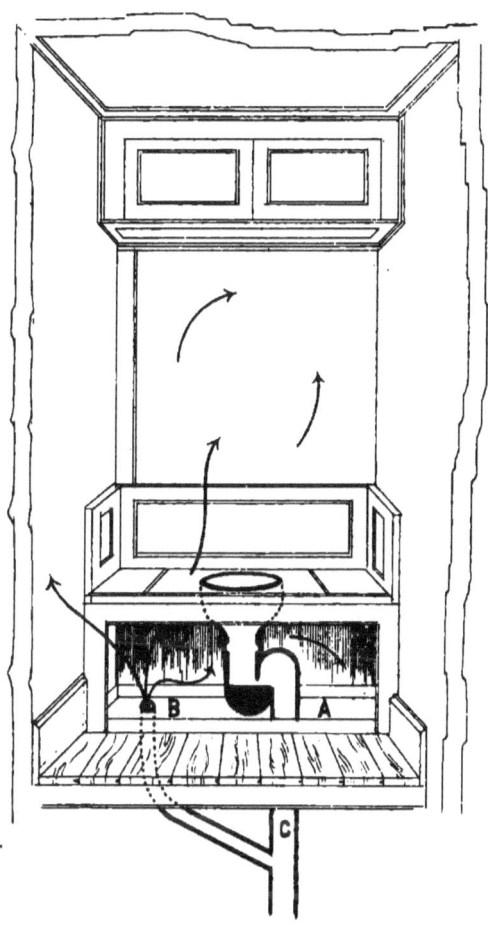

"Save-all" tray beneath w.c. with untrapped waste pipe acting as ventilator of soil-pipe.

PLATE XXIII

Soil-pipes inside a house

Are rarely safe, and still more rarely excusable. Sometimes they are "let into" the wall and covered by a board, and perhaps papered over, and thereby concealed. The board acts as a shaft to conduct any leakages of sewer gas into the upper rooms.

The sketch for this drawing was given me by a medical student who occupied the room in question. He was constantly suffering from sore throat, and after repeated accusations of the drains, and with much reluctance, his father at last allowed an investigation. The concealed soil pipe was discovered in the wall of his study, perforated at several points, and ending below the floor in a broken junction with the drain.

Sometimes a leaking soil-pipe is discovered in the wall of a kitchen cupboard where food is kept.

PLATE XXIII.

"Soil pipe" in corner of sitting-room, concealed by a board.

PLATE XXIV.

Leaden soil pipe, seamed, and crumbling with age.

This was found in a house recently occupied by a relative of my own. An old water-closet, very little used, and situated in the centre of the house, was condemned to removal. The plumber who removed it found the soil pipe so rotten that it "crumbled like short cake." It was open at the seam, so that not only gas but liquid sewage had escaped, and had made the contiguous wall, and the kitchen under which the w.c. drain ran, "black damp."

The soil pipe of a w.c., if inside a house (an arrangement better avoided), ought to be made of *drawn lead*, i.e., *not* of sheet lead rolled into the form of a tube and soldered at the seam. A seamed pipe may be defective, and leak at any point of the seam. A drawn lead pipe, if a good one, is only in danger of being defective at the joints, *vide* Plate XLV., "Putty Joints."

Age is a source of danger in leaden soil pipes. Dr. Fergus, of Glasgow, found that unventilated pipes of 15 years, and ventilated pipes of 25 years, became eroded, eaten into holes, on the inner surface, by the sewer gases, especially on the upper surface of a bend. Dr. Fergus considers that the duration of *ventilated* soil pipes is from eighteen to thirty or more years; of *unventilated* soil pipes from a minimum of eight years to a maximum of twenty years.

He has traced illness on many occasions to perforation from within of leaden soil pipes, which had been corroded by sewer gases.

PLATE XXIV.

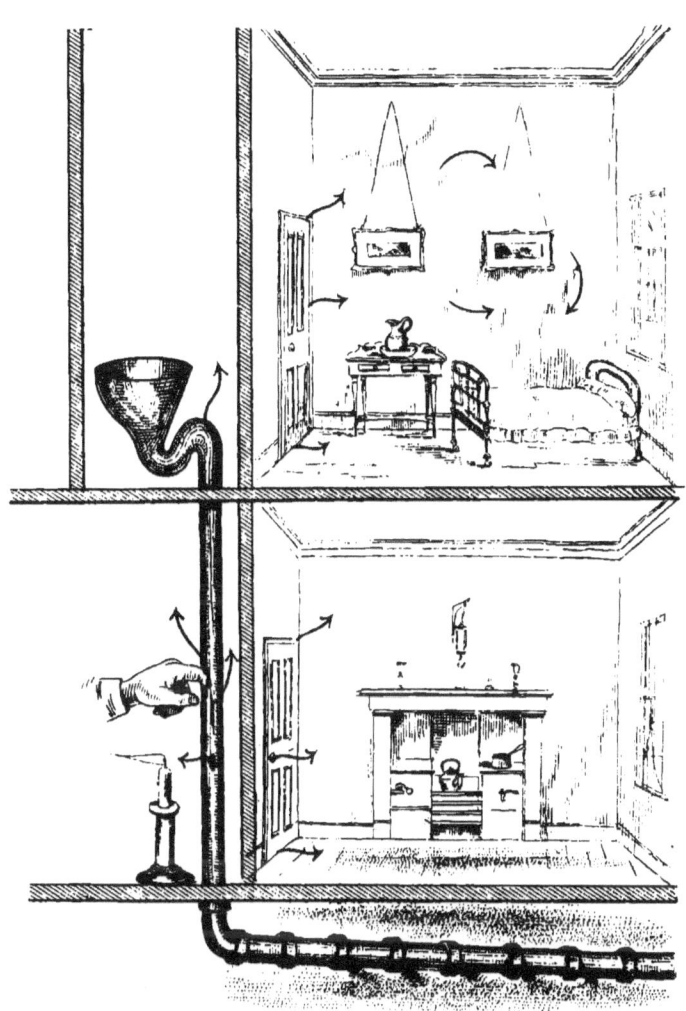

Leaden soil pipe, seamed, and crumbling from age.

PLATE XXV.

Scullery sink discharging over the grate guarding an untrapped drain.

This drawing was contributed by Mr. Chorley, who discovered the defect in the house of a relative.

Mr. C. had noticed a drain smell in the hall and lower part of the house. On investigation, he found the sink pipe (A) delivering its waste water into a grate (B) which covered a sinkstone of an untrapped drain. This drain joined a w.c. drain running under the house. In the same house he found an untrapped sinkstone in the "keeping cellar." Before these faults were discovered and remedied the lady of the house was constantly in ill-health. Since the correction of the faults her health has been perfectly restored.

PLATE XXV.

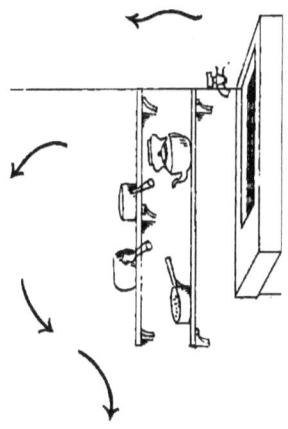

PLATE XXVI.

"No wonder the meat wont keep, the beer turns sour, and the milk disagrees."

"Dish-stone in larder leading into a drain."

Open grates in cellars for the purpose of "swilling" the floor are not uncommon. They are often untrapped, and when trapped, the traps are usually ineffective from want of water, or from being broken; and even if sealed by water, they are still an inefficient barrier to sewer gases, which can pass by absorption through water.

In the dairy and larders of the new Leeds Infirmary there were found sinkstones practically untrapped in every instance.

It is probable that this communication with the drains may have been the explanation of certain outbreaks of diarrhœa in the Hospital which were attributed to the milk, but without any such source of its contamination being suspected.

About 3 years ago two boys were ill in low fever in a newly-built country house. Every care had been taken about the drainage, and the drinking water was found free from pollution. The medical attendants were for a long time at a loss to find out the source of the fever. At last the milk was suspected, and the dairy at some distance from the house was examined. A sink-stone leading to an open drain was discovered.

PLATE XXVI.

Bad keeping Cellar.
"No wonder the meat wont keep, the beer turns sour, and the milk disagrees."

PLATE XXVII

"Dairy Sweepings."

This illustration was contributed by Dr. Midgley Cockroft, of Masham, in the following letter :—
"I attended the family on two occasions. In the first the "type was purely Typhus,—four cases, one death. On the "second attendance the type was entirely Typhoid; all had "diarrhœa, all had rose spots, and one death occurred in the "four cases, three of the cases having gone through the first "illness. There was no other case of either variety in the "neighbourhood. I had a good opportunity of watching the "process of cleaning down the dairy. The joints in the "flagging were purposely left about $\frac{3}{4}$ of an inch apart in "order that the water thrown on could easily be *brushed into* "*the fissures*, whence I could hear it falling into a drain below, "which drain only went from the dairy into a garden in front "of the house, a distance of about 10 or 12 yards, with a very "little fall in its course. The house was a very old one, and "has now been replaced by an entirely new one. I may add "that the dairy floor was 'dished' to facilitate the discharge "of the water." It would seem that the spilt milk, washed into the imperfect drain, underwent a poisonous decomposition in the drain, and thus gave off poison to the dairy, milk, and kitchen.

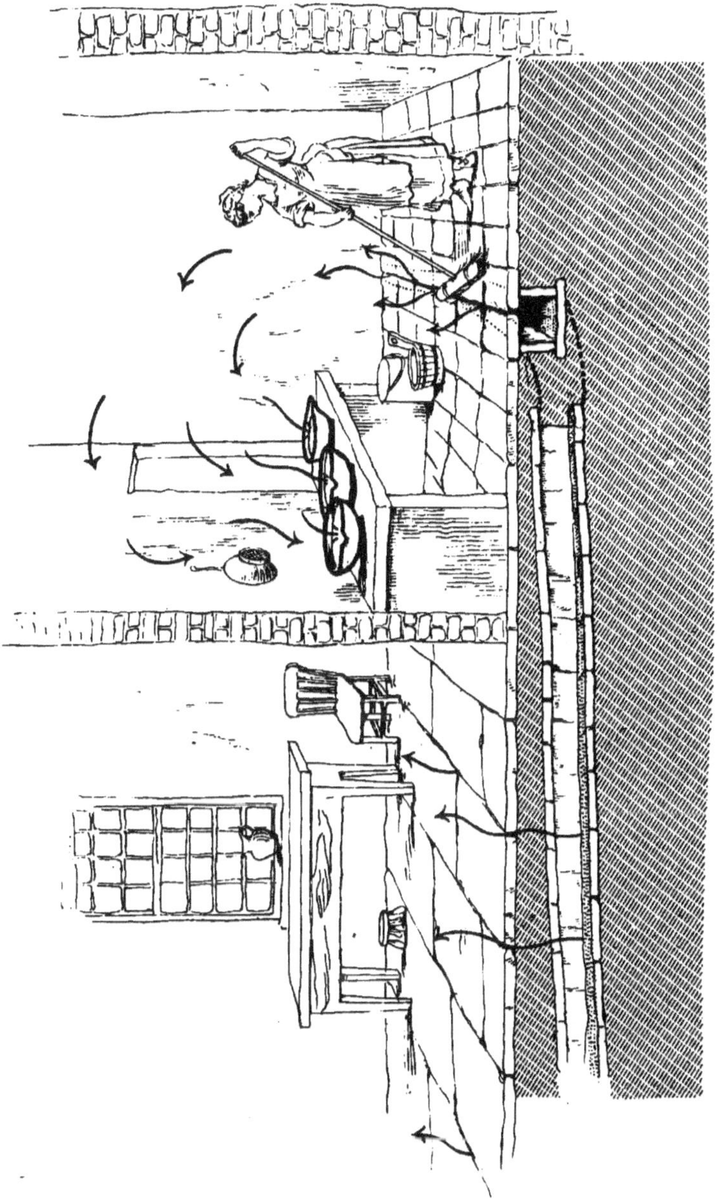

PLATE XXVII.

Dairy Sweepings.

PLATE XXVIII.

"Dish-stone" in scullery leading into a rain-water tank with overflow direct into a drain.

This illustration, as well as Plates VI. and XXIV., was taken from the house of my relative. The servants were in the habit of washing the floor, and sweeping the "washings" through the sink-stone into the tank. The tank had an overflow direct into the drain, and thus the sewer air had a free passage into the house.

PLATE XXVIII.

Dish-stone in scullery untrapped, and opening direct into a rain-water tank, with overflow into drain.

PLATE XXIX.

Pantry sink turned into soft-water cistern under the cellar floor, with overflow into the rock.

Contributed by Mr. Edward Atkinson, from his own house.

A large soft-water cistern was discovered under a cellar floor, full of very offensive water, which, having no overflow pipe, must have overflowed into the foundations. The cellars had been excessively damp, and had baffled costly attempts by his predecessor to remedy. Into this tank the slops from the butler's pantry found their way, as the waste-pipe of the sink had been turned into an old channel under the cellar floor which conducted rain from the fall-pipes into the tank. The butler's sink was one of the improvements preparatory to Mr. Atkinson's purchase.

A similar fault was communicated to me by the late Mr. William Gray, of York. "I send you a sketch from our "court which I think equals any you have recorded. A new "sink being put down, it was obviously a *short cut* to discharge "it into the *fall pipe*, which fed a tank under the sitting-room "floor. Two successive families occupying the house had "typhoid fever."

PLATE XXIX.

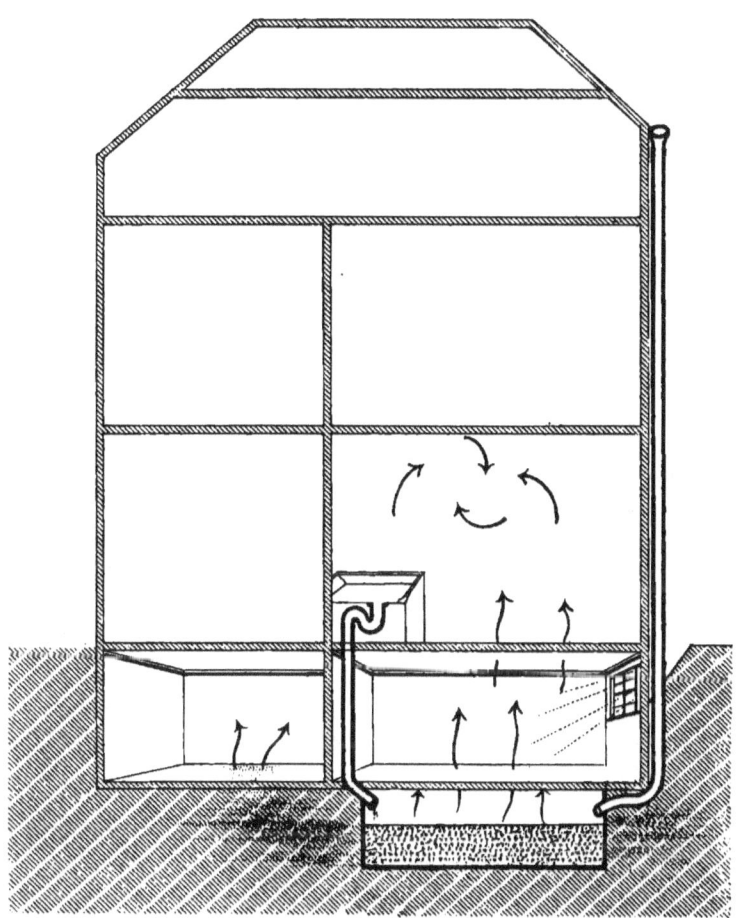

Pantry sink turned into soft-water cistern under the cellar floor, with overflow into the rock.

PLATE XXX.

Disused and unsuspected water-tank under cellar-floor.

This was found in the house of Mr. H. B. Hewetson, surgeon, of Leeds. Sometime previously, in consequence of illness in his family, he had removed a central w.c. to the outside, and had, as far as he could judge, corrected all sanitary defects. Illness of a typhoid character broke out, affecting Mr. H. himself and a maid servant. This led to a search under the cellar steps where the flags sounded hollow. A large unsuspected tank (A) was found, with direct overflow into the drain (B). The end of a pipe of a long disused water-closet (C) was discovered at one corner. At the same time some other defects were found and remedied. Mr. H. recovered after a few days' illness, and the servant lingered four months and then died.

The workman (D) is inserted in order to shew how spaces under stone floors can be discovered by the hollow sound produced by a falling crowbar.

PLATE XXX.

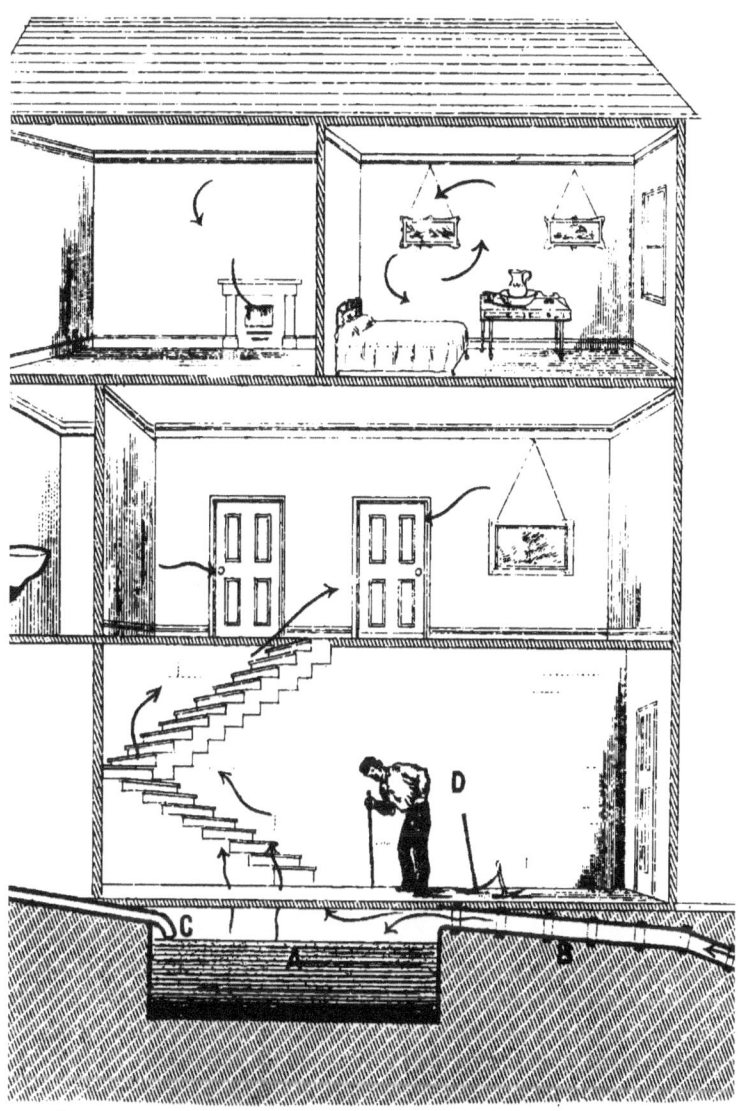

A. Rain-water tank under cellar floor, with overflow into drain.
D. Workmen "sounding with crowbar" for suspected "tank" or "cesspool."

PLATE XXXI.

Rain-water cisterns and the dangers they entail.

The fault here presented is one not uncommon in country houses.

A large cistern (A) within the house receives the rain water from the roof, and, as a matter of course, has an over-flow pipe (B) to carry off surplus water.

The over-flow pipe (whether it be trapped or not is of no moment or value) conveys the water into a large storage tank (C) outside.

The storage tank again has an over-flow into a cesspool (D). The cesspool is sealed at the top and unventilated. The gases formed in the cesspool and drains pass backwards along the overflow pipe into the tank, and thence along the first overflow pipe of the cistern into the roof of the house, whence they are drawn by the house fires into the rooms and passages.

PLATE XXXI.

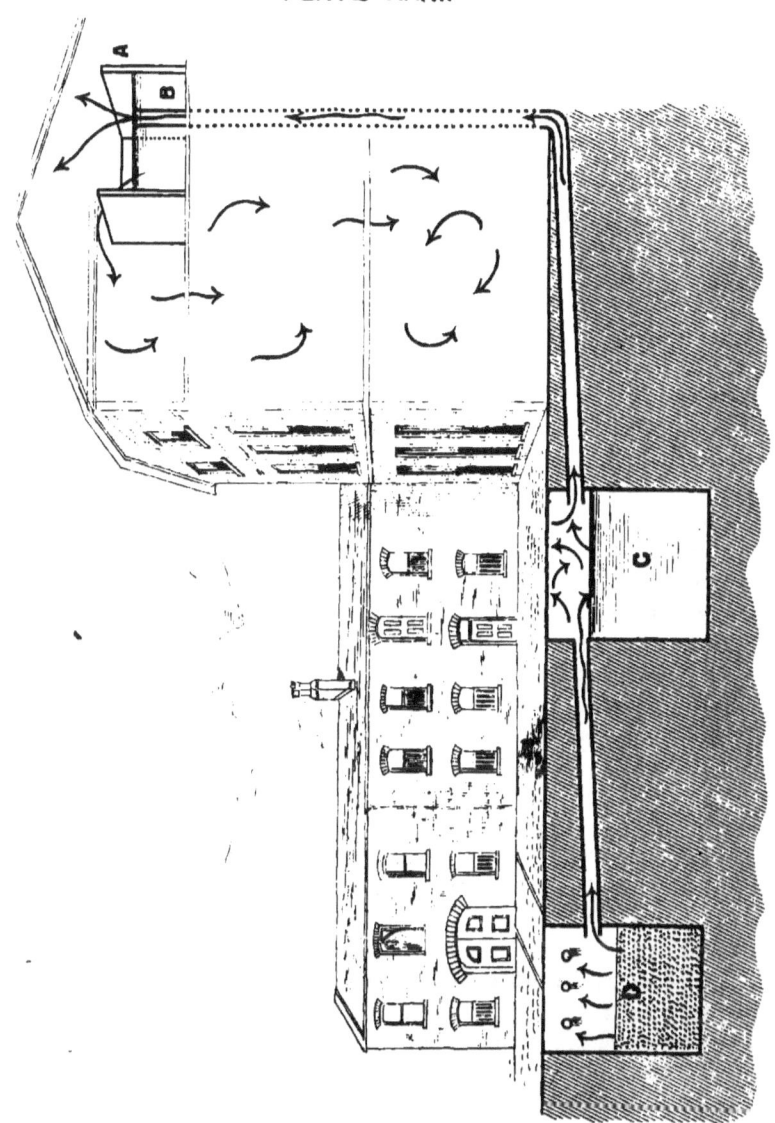

How cesspool poison may steal into a country house.

PLATE XXXII.

How people drink sewage.—No. I.

Drain pipes badly joined or broken, leaking into a well.

This is the condition probably of a large proportion of the wells of the country, especially of the shallow surface wells.

A glance at the picture will convince most thinking persons of the pressing need there is for a great national organisation for providing wholesome drinking water to villages and small towns which do not as yet possess a public unpolluted water supply. This need has been pressed upon the attention of the public by the press, the Society of Arts, and by His Royal Highness the Prince of Wales.

A well may be polluted with sewage for a long time before illness results.

The history of an outbreak of typhoid fever at a large school about *ten years ago* is almost classical. By judicious care and outlay the health of the boys in this school had been long preserved at a high level. But on the reassembling after the holidays a boy fell ill of typhoid fever contracted at home. He was placed in the "sick-house," and used the w.c. which discharges into a drain running near the underground cistern which supplied the drinking water. In a fortnight about 30 boys were down with typhoid. A careful investigation made by the proprietor and Mr. Ellerton, and reported on to the Local Government Board by Dr. Clifford Allbutt, revealed a leakage from the drain, and a fouling of the cistern thereby. Both cistern and drain had been very carefully and properly constructed, but the drain lay too near the cistern, so that when a joint of the drain was let down by a rat run, the escaping sewage soaked through some fine crevices in the cement of the cistern.

In this instance, water fouled by drainage did not set up typhoid fever until the importation of a case of typhoid led to the introduction of typhoid discharges into the drinking water.

PLATE XXXII.

How people drink sewage.—No. I.
Drain leaking into a well.

PLATE XXXIII.

How people drink sewage.—No. 2.

Cesspool full and overflowing into a well.

This is the same in principle with the last picture, and teaches that cesspools need constant attention and cleansing, and very great care in construction. (Building Bye-laws, § 35—§ 39.)

The following illustration came before my notice :—Typhoid fever broke out at a farm house, distant about half-a-mile from a village. The father died, and the mother and daughter recovered. In the village the only case of fever which occurred was that of the farming man who had his meals at the farm and went home to the village to sleep.

The following was supposed to be the cause of the fever :— Ten feet from the door of the house there was a cesspool. Twelve months previously a well had been made between the house and the cesspool. Shortly before the time of the fever the house drain had become so offensive that the cesspool was examined, and was found to be full and overflowing, and was in consequence emptied. It is a fair inference that the cesspool had overflowed into the well and poisoned the drinking water. This occurred some years ago, and the well-water was not analysed, so that complete proof was absent.

Nevertheless the picture generalises well known and well ascertained facts.

I have recently seen in Leeds, during the excavations for a new building, the exact conditions here depicted. The well was formed of bricks laid radiating from the centre of the well, and of course with plenty of room between them for admitting leakage into the well from the neighbouring cesspool.

PLATE XXXIII.

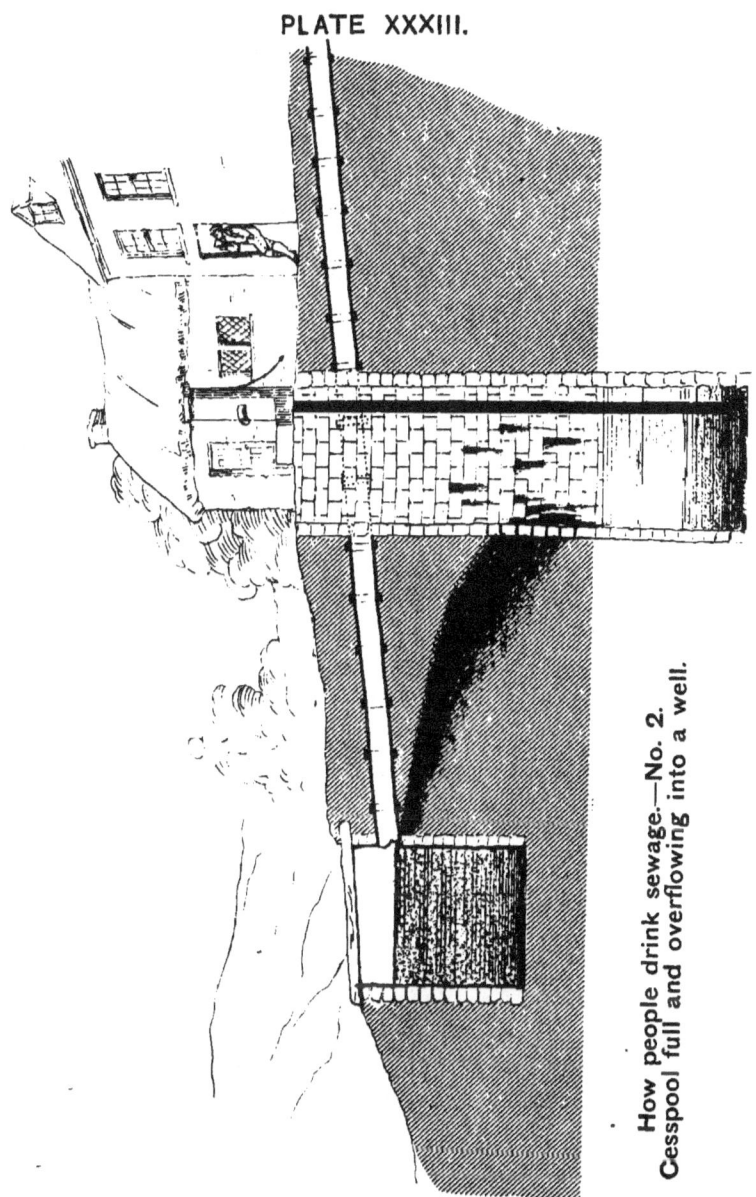

How people drink sewage.—No. 2.
Cesspool full and overflowing into a well.

PLATE XXXIV.

How people drink sewage.—No. 3.

Well near "fold yard."—Drain carried over a well.

This picture represents a series of facts.

No. 1.—That wells are often placed within, or close by a farm yard, so that the soakings from the sodden manure must needs ooze through the soil into the well.

No. 2.—That, inconceivable though it may appear, it is a fact that drains are sometimes carried across the upper part of a well. Two instances have been related to me.

The first by Mr. Robert Hagyard, late a student at the Leeds School of Medicine :—

"A drain ran through the side of a well, the pipes projecting so as to be visible when the lid was removed from the top of the well. At the junction of the pipes a leakage of sewage, consisting chiefly of liquid from a pigstye and cowshed, trickled down the side of the well. Several cases of typhoid fever had occurred in the house. The occupants sold milk, which was freely diluted with the 'solution of enteric fever' from the well."

The second was related to me by Mr. John Bradley, coach builder, of Leeds :—

"The drain is carried over the mouth of a well on a plank, which rots and lets down the drain, and pours the sewage matter into the well which supplies a large establishment with drinking stuff. Typhoid fever broke out, and one of six cases ended fatally, when the above state of things was discovered."

No 3.—That milk is made poisonous by the use of contaminated well water, either for diluting the milk or even for washing the cans.

Moral.—Every dairy whence milk is sold ought by law to be under constant sanitary inspection.

PLATE XXXIV.

"The 'well' contrived a double debt to pay." Well and cesspool all in one.

PLATE XXXV.

Overflow from cesspool into rain-water tank.

This plate represents a fact related to me by my friend, Mr. W. P. Goodall, surgeon, of Birmingham.

In a newly-built vicarage a rain-water tank had an overflow pipe into a cesspool, the levels of which were so skilfully mismanaged, that the cesspool, when full, relieved itself by overflowing back into the rain-water tank.

A second fact of the same kind is thus told to me by Mr. John Bradley, of Leeds:—

"A new bank, with residence for the manager, was erected in a small market town. Shortly after he went to reside there his wife became ill. She went from home for a week, and returned quite well, but found her servants and children attacked as she had been. At this time a great stench was felt in the scullery near the pump of a soft water tank. The tank was examined, and was found three parts filled with sewage. The builder had laid the overflow pipe into the sewer, with the fall the wrong way, and had thus tapped the sewer, and the sewage had flowed into the tank."

PLATE XXXV.

A New Vicarage. Cesspool overflowing into a tank.

PLATE XXXVI.

Cesspool overflowing and causing the floor and wall of a house to be damp from sewage.

This illustration is a general expression of the following facts, rather than a representation of any actual example.

Case 1 was related to me by Dr. James Braithwaite, as having occurred in a suburb of Leeds about four years ago.

Typhoid fever occurred in two of a group of three newly built houses, within a few weeks of their being occupied. The following conditions were discovered when, *a few months after*, the main drain was brought within reach, and an attempt was made to connect them. The drains from Nos. 1 and 2 opened into the drain of No. 3, and this terminated 18 inches from the house, forming a cesspool in the soil which rested against the cellar of No. 3, and in rainy weather caused the cellar floor to be flooded. Typhoid fever broke out in No. 3, and afterwards spread to No. 1.

Case 2.—A young woman was suffering from chronic sore throat and partial loss of voice, a serious matter, as she was being trained as a public singer. Having enquired into the sanitary condition of her house, I learnt from her mother that two children had died of diphtheria, and that the kitchen floor was damp and offensive from the overflow of their cesspool. Complaints had been ineffectually made to the landlord's agent, but her husband dared not complain to the landlord, his master, for fear of being dismissed from his situation, that of head gardener.

PLATE XXXVI.

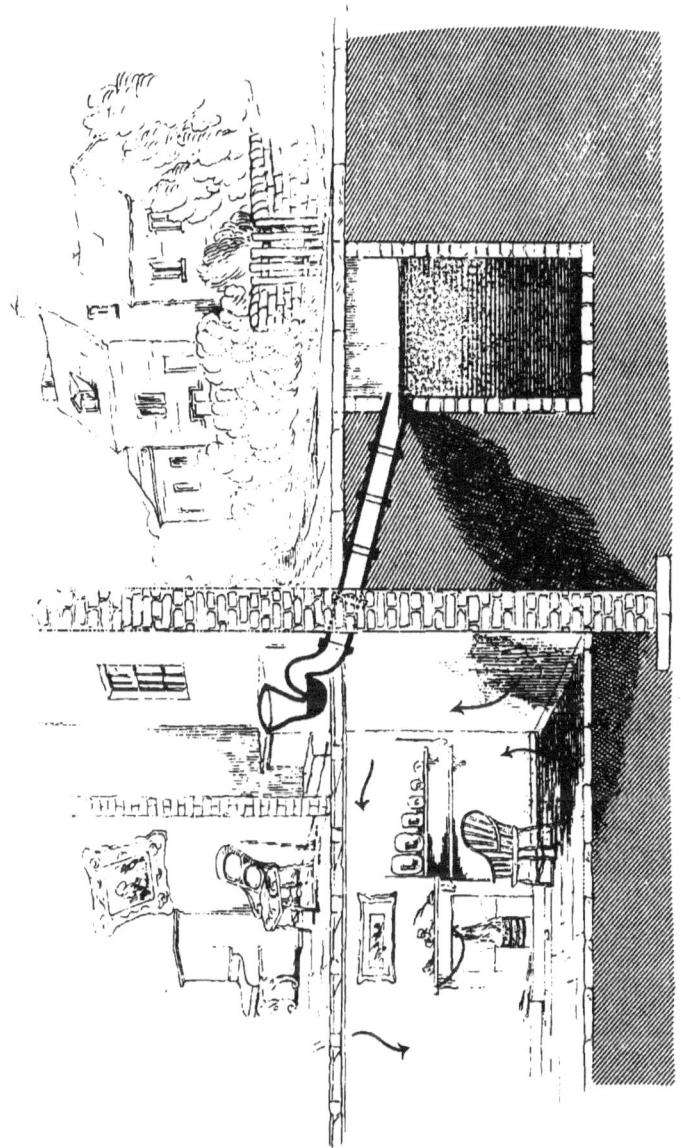

Dampness of house from overflow of cesspool.

PLATE XXXVII.

Additions to house built over forgotten drains.

For this illustration I am indebted to Dr. Britton, of Halifax.

A billiard room was built on a vacant space between a house and a stable.

After a time the billiard room was converted into a dining room.

Typhoid fever broke out two weeks after the family returned from the sea side. A child was ill and recovered, and a servant died. This led to a sanitary inspection of the house, and the discovery that the waste pipe of the kitchen sink joined an old drain which led to a cesspool under the new dining room, the existence of which had been previously unsuspected.

Moral.—In adding to a house, make sure that all drains traversing the site and all cesspools have been obliterated.

PLATE XXXVII.

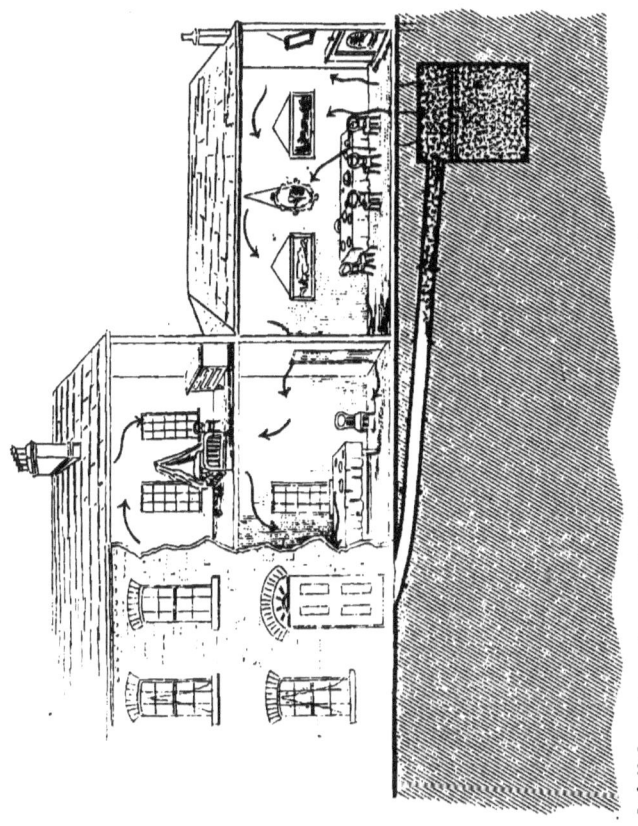

Addition to a house built over an unsuspected cesspool.

PLATE XXXVIII.

"Where is the Butler?"

For this fact I am again indebted to Mr. John Bradley, of Leeds.

"A gentleman came to reside in an old family mansion. Having friends to dinner one evening, and requiring more wine, he rang the bell. No butler came. He rang a second and third time with the same result. Waxing wroth, he went in search, but could get no tidings of him. On further search the butler was discovered in an old cesspool in the wine cellar, the floor of which had broken in. The poor butler, after much difficulty, was extricated, only just in time for his life to be saved after much suffering and a month of medical attendance. It appears that the existence of this cesspool was unknown, and that so long as the sewage of the house went 'somewhere,' no enquiries were made to ascertain where."

A similar fact was told to me by a lady, as having occurred in a large house at Brighton. A cask of beer was being rolled along the cellar, when the floor gave way over an unsuspected cesspool.

Moral.—Test all the floors of your cellars by "sounding." *Vide* Plate XXX.

More wine wanted. Where *is* the butler?

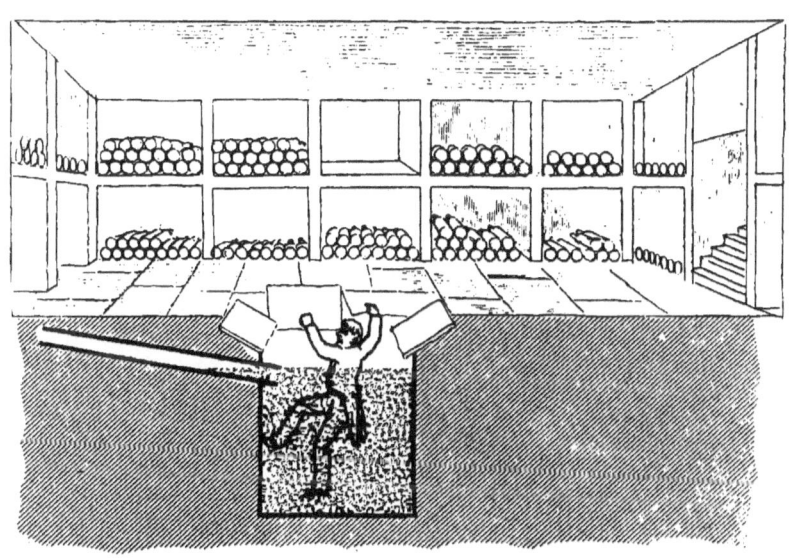

"A lower deep,
Still threatening to devour me opens wide."—Milton.

PLATE XXXIX.

Broken junction of drain with soil-pipe, leakage into disused well under keeping cellar.

This fault was discovered in a house in Park Row, formerly occupied by myself, but now used as offices. About three years and a half ago complaints were made of bad smells in the house, and some of the inmates were unwell. On inspection it was discovered that an old disused well partly under the keeping cellar was becoming a cesspool from leakage through its walls from the w.c. drain. This drain had become defective at the junction of the vertical soil-pipe with the horizontal drain. It appeared that the *soil-pipe had settled, and by its weight had broken the flange of the drain-pipe*, causing the sewage to flow into the rock underneath the cellar floor, and so into the well. The drain pipes were repaired, and the well was filled up.

PLATE XXXIX.

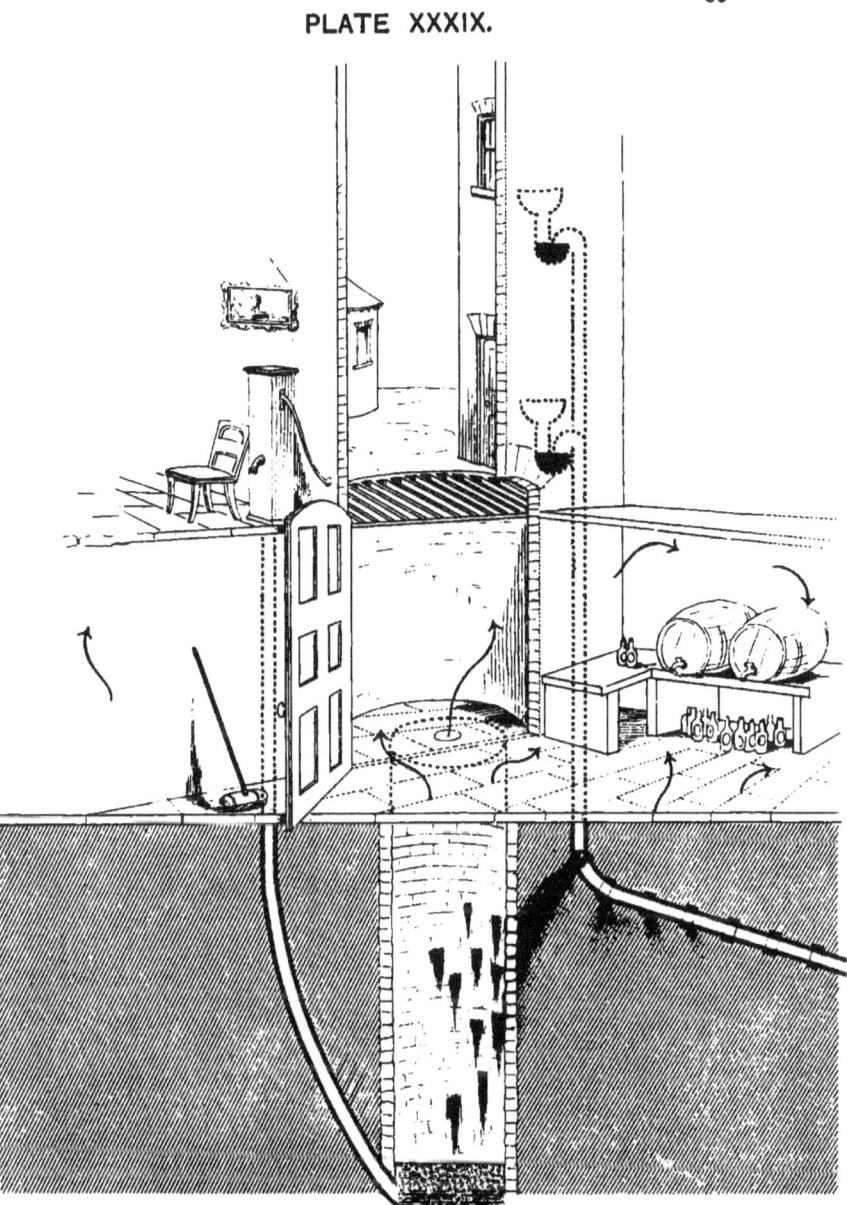

Well under a house fouled by leakage from broken junction of soil pipe with drain.

PLATE XL.

Common stone drain under tiled entrance hall, leaking at every joint, and forming an extensive cesspool under the house.

This example was communicated to me by Dr. Britton, medical officer of health for the Halifax Rural Sanitary district, in the following note :—

"Enteric (typhoid) fever broke out in a gentleman's house, "from which it spread into the village. On examination I "found that the w.c. was *in the centre of the house*, and that "the soil-pipe discharged into a common stone drain running "under a *tiled* entrance hall. This drain was almost without "fall, so much so, that it had become blocked, and the *sewage* "*had found its way under the flooring of the passage and rooms.*"

It goes to a man's heart to take up a tiled hall in order to inspect a drain. *Moral.*—The drain ought never to have been placed under the hall.

PLATE XL.

Common stone drain under tiled hall, leaking at every joint, and forming a large cesspool under the house.

PLATE XLI.

Joints opened by giving way of foundations.

"When drains are laid in new made ground, unless care be
"taken to ram the earth sufficiently hard round about them,
"and this is next to impossible, the pipes will open at the
"sockets, and sodden the ground in their neighbourhood to a
"dangerous extent."
Sanitary arrangements of Dwellings, Eassie, p. 22.

This may occur in laying drains in newly-made ground, and it frequently does occur where the drain trench has been unevenly cut, and where inequalities of level are carelessly filled in with soft soil, which after a short time settles, and allows the joints to open.

PLATE XLI.

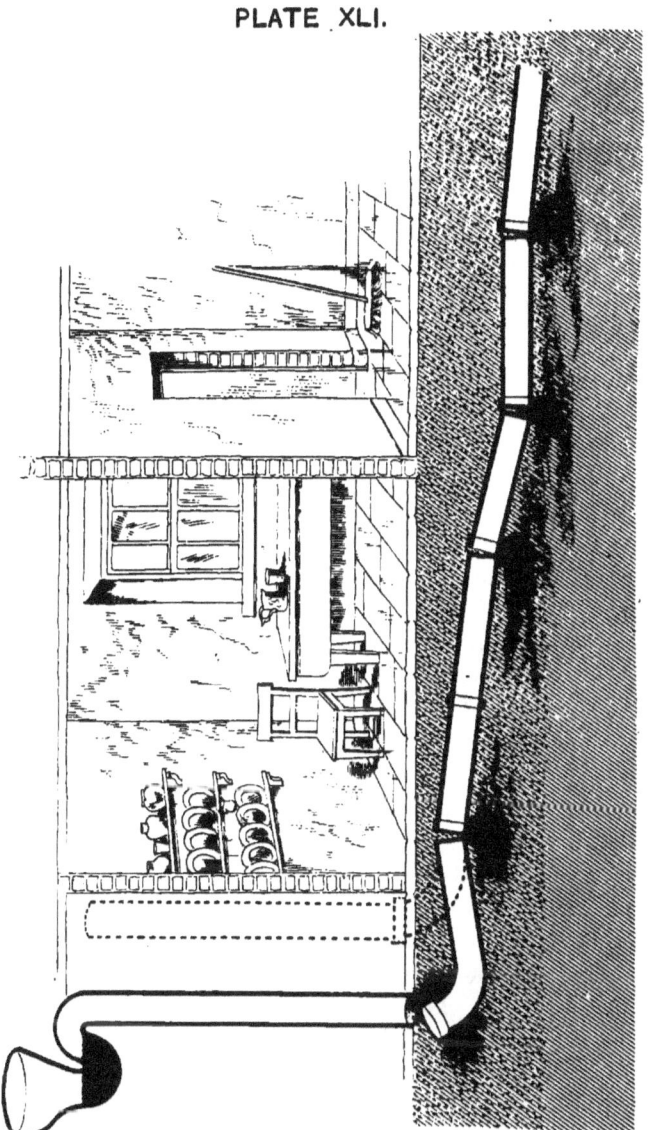

Joints gaping from sinking of foundations.

PLATE XLII.

"Poisoned by next door neighbour's drains."

It is not easy to obtain an unexceptionable illustration of this danger. I feel convinced, however, that it does occur occasionally, either from leakage of drains soaking the party wall, or from a neighbour's soil-pipe running in the thickness of, or even on the inside of the party wall of the suffering house, or again from the *diffusion of sewer gases through the wall itself*.

Since the publication of the first edition of this book several instances of this danger have been communicated to me.

For the drawing and facts of this plate, I am indebted to Mr. Foster, Artist, of Headingley, who says—" I enclose you a
" sketch of the defect in the drainage of the house in which the
" fatal case of typhoid fever occurred. The foundation of the
" house had given way, and the earth at the side sinking with
" it opened the joints of the pipes which ran under the yard of
" their neighbour's house."

It is difficult enough to manage one's own drains, almost Utopian to hope to rectify the drains of one's neighbour.

PLATE XLII.

"Danger from next door neighbour's drains.
Proximus ardet Ucalegon."—Virg. Æn. II. 312.

PLATE XLIII.

Speculating builder buying "Seconds."

On one of the occasions of the delivery of my lecture in a suburb of Leeds, one of our leading builders stated that it was well known by the building trade that dishonest builders of cheap houses were in the habit of buying "Seconds" sanitary tubes, *i.e.*, rejected broken tubes, at half price, in order to lay them in the houses they were building, in obedience to the law requiring them to lay a drain. Such tubes are defective either by fracture or by being mis-shapen, oval instead of round, or *vice versa*. Each such defect would allow a leakage, and the formation of a cesspool at the faulty point. In drains, as in chains, the value of the whole drain is determined by the value of its weakest point, and if at the weakest point there is a leakage, the whole drain may be worthless and disastrous.

If this picture has the effect of gibbeting such scoundrels, and making scamped drain-work less feasible, it will have served its purpose.

"'Jeremiahs' buy 'seconds' because they can't get "'thirds,'" said an honest Yorkshireman on seeing this picture.

"Jerry veal" is the flesh of calves which have been born dead, or have died soon after birth—an "article of commerce" in former days.

PLATE XLIII.

"Jerry builder" buying "Seconds."

PLATE XLIV.

Drain made of "Seconds" tubes.

Here are seen the results of scamped drain-work and cheap "Seconds" pipes. Such pipes are used mostly for the outside drains of cheaply built cottages and houses, and are sometimes found inside a house.

The pipes AA. are broken at the flange, BB. at the smaller end, and FF. are mis-shapen, spoiled in the baking, oval instead of round. Each of these defects renders a sound joint impossible.

C. has a fissured surface, D. has been broken and pieced together, a condition of pipe which Mr. Burton, the lithographer of this book, himself witnessed in his own house, and which he has drawn '*con amore*.' The workman declared that he could not afford to put in a new pipe.

G. shews careless connection of a waste-pipe. Instead of a tube with a proper junction as part of its construction, a hole has been broken into the tube, and the lead pipe passed through without luting. Moreover, the waste-pipe projects so far into the drain-pipe as to form an obstruction to the proper flow of sewage. *Vide* Plate XLVII.

A drain formed of imperfect tubes with unluted joints, and insufficient fall, was found under the house of Mr. Carter, dentist, in Park Square. The soil under the floor of the kitchen was saturated with sewage, and the villany was rendered complete by the entire omission of a pipe for connecting the drain with the main-sewer.

Mr. Carter, by his removal into this house, got "out of the frying-pan into the fire." He had left his previous residence in consequence of "drain-begotten" illness in his family, and because of the rats which he had shot with an air gun by the dozen in his kitchen.

A medical friend illustrated one of my lectures by "seconds pipes" just discovered in his own house, which had been recently built at a cost of £3,000.

PLATE XLIV. 90

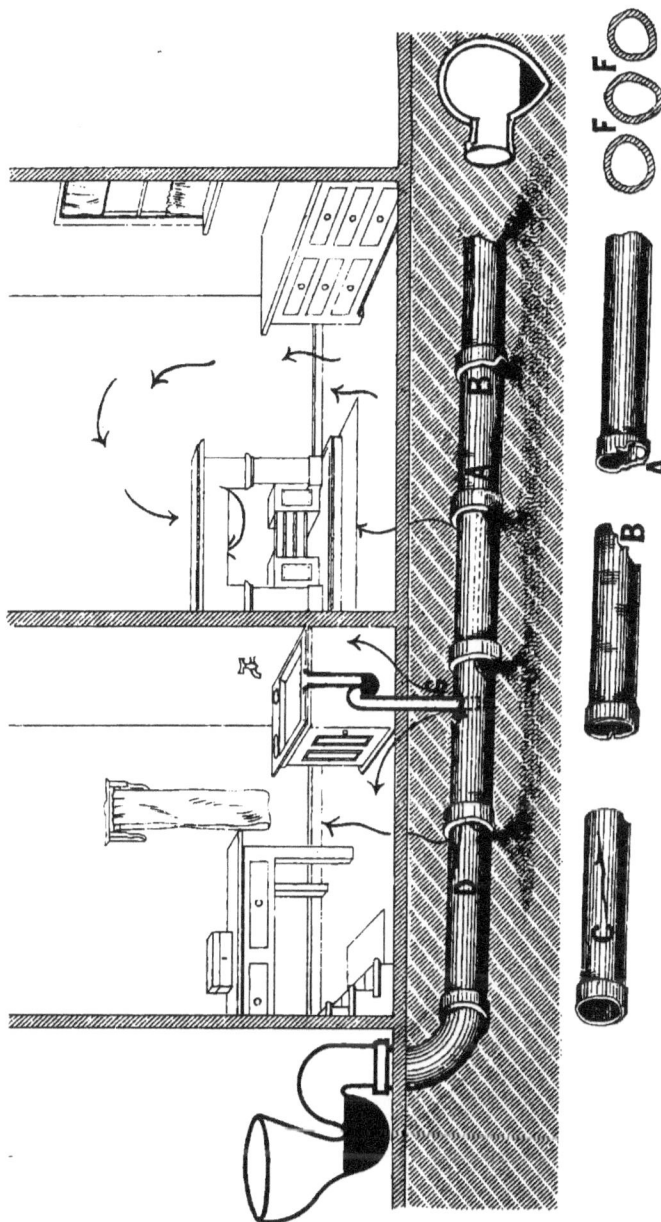

Drain made of "Seconds." Manslaughter under an "alias."

PLATE XLV.

"Putty joints" in leaden soil pipes.

This is scamped work. In order to save his pocket the plumber will sometimes save the cost of solder, and join the leaden soil pipes with putty and inferior material. The result is that the joint is insecure, soon gives way, cracks and gapes, and allows sewer gas to escape into the house.

A flaw in the joint can be detected by the current of air against the flame of a candle, and the quality of the material may be tested by its easily giving way to the finger or a knife.

Leaden soil pipes ought to be carefully joined together by solder, and to have no crevice through which air can pass.

PLATE XLV.

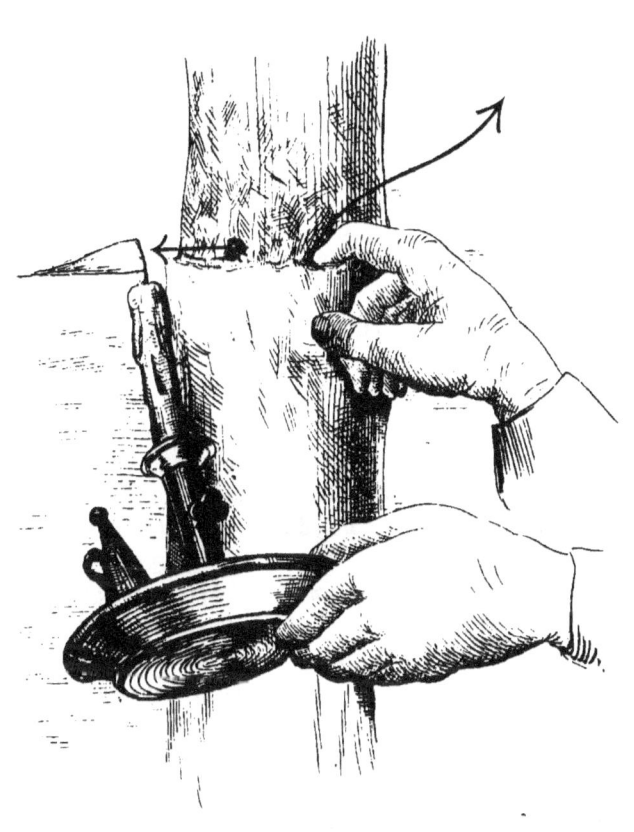

"Putty Joints."

PLATE XLVI.

Curves made by straight pipes.

Such work is down-right scamping. To save trouble or expense straight pipes are joined at angles which allow gaping and leakage at every joint. If such ill-made bends are under a house, a large cesspool is there formed—if outside the house the leakage may soak towards the wall of the house and make it damp, foul, and poisonous.

About two years ago a friend of mine was preparing his house for his intended bride, and like a wise man had the drainage looked to, and finding it very bad, was re-constructing it. He took me early one morning to see the new work that was being done, and to show me a bend made by straight pipes, part of which we had to uncover in order to see it.

Another instance was found in the house of a patient of Mr. Horsfall, of Leeds, who, in consequence of the illness for which he was in attendance, urged an exploration of the drains. This examination discovered a drain under the kitchen floor, not only open at the joints, but with bends made by straight pipes, so that half the sewage had remained under the floor.

PLATE XLVI.

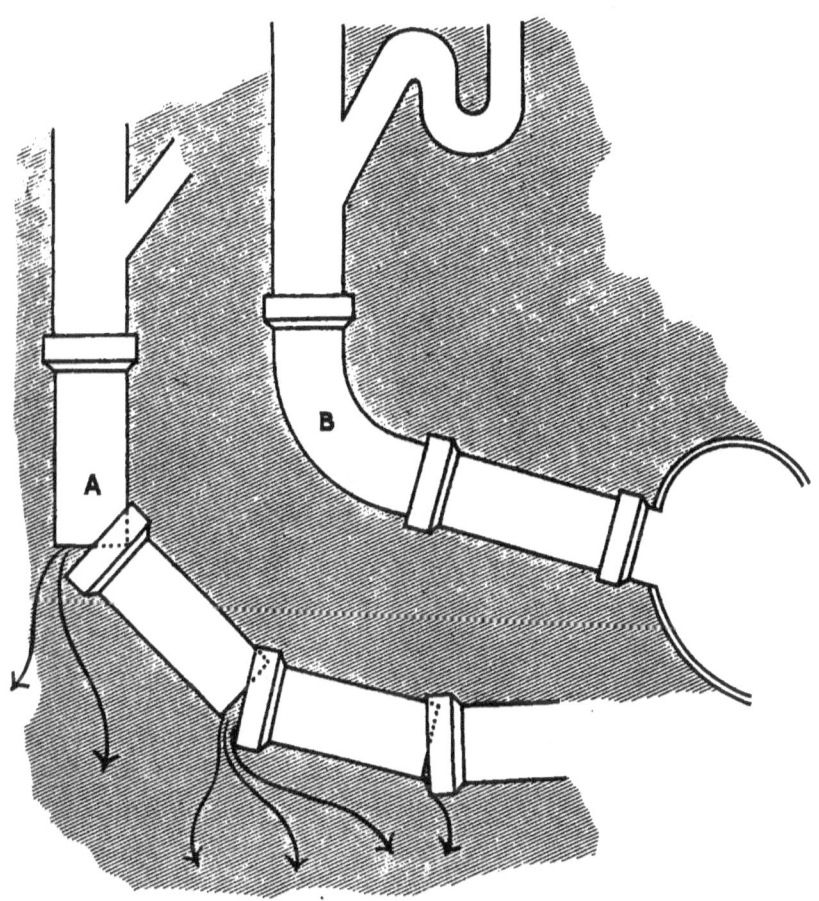

A. Curves made by straight pipes, leakage at every joint.
B. Curve made by proper bend.

PLATE XLVII.

"Junctions."

A. This is scamped work. From idleness or false economy a junction is made with a drain pipe by breaking a hole at the top or side of a pipe, and simply passing the joining pipe through the hole. This involves at least two grievous defects—(a) the hole through which the joining pipe passes cannot be properly luted—(b) the intruding pipe projects into the receiving pipe, intercepts the solid parts of the sewage, and by degrees dams it up.

Several instances of this fault have been related to me—one by the medical officer of a recently-built workhouse in Yorkshire, in which, in addition, it was found that the main drains were led "up hill;" a second was communicated to me by Dr. Churton, of Leeds, as having been discovered in a house he recently occupied.

B. A properly-made junction.

C. Sometimes it is necessary to tap a drain and let in a new pipe. In such a case the hole carefully cut in the old pipe should be capped by an "eyelet," and made safe by cement.

PLATE XLVII.

A. Badly made junction. Blocked drain.

B. Proper junction.

C. "Eyelet" for making a new junction in pipes already laid.

PLATE XLVIII.

Pipes laid the wrong way.

This arrangement of pipes was discovered in our new Infirmary, by Mr. Chorley. Rain-fall pipes carried under a room, were leaking at each socket, rendering the soil damp.

This arrangement reduces very greatly the "water tightness" of the joints.

I have been told that builders in some parts of Yorkshire maintain that to place drain-pipes upside down is the correct way.

PLATE XLVIII.

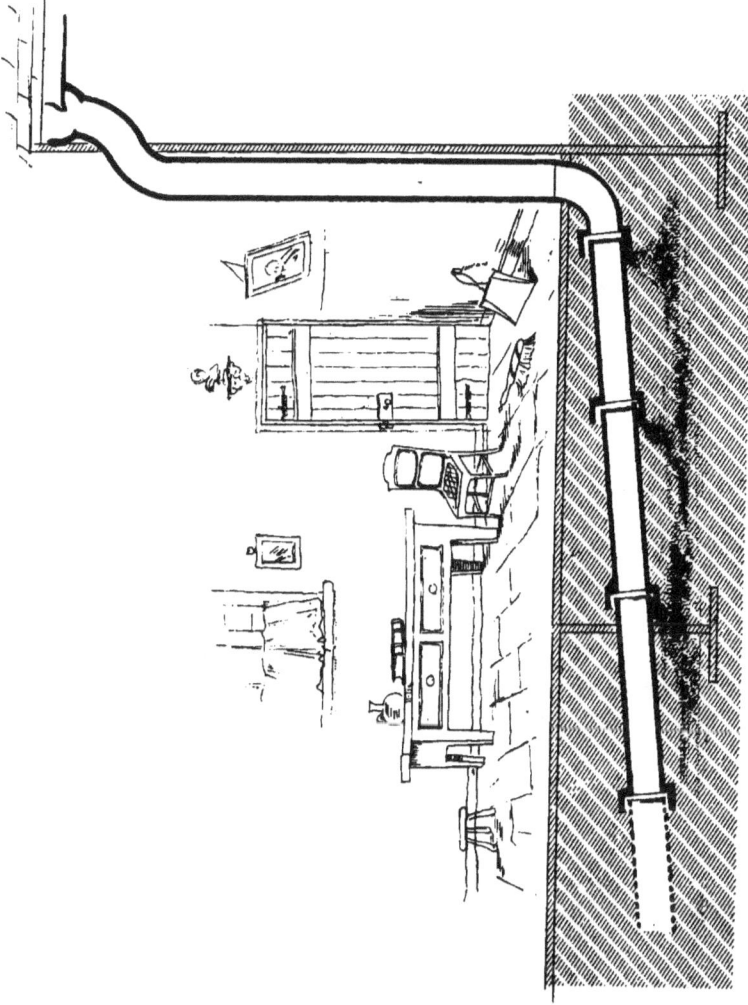

PLATE LIX.

Drain under a house running up-hill.

This illustration is contributed by Mr. Pickles, Surgeon, of Leeds. He had a slight scratch on the finger from which inflammation started and spread up the arm, due as he supposed to poison received in attending a patient. Soon after the recovery of the arm, he was again laid up with rheumatism of a low type. His medical attendant suggested that the house drain was probably the cause of the whole mischief. As soon as he was well enough he had his drains examined, and reported the result to me in the following note.

"I have had all my drain-pipes taken up, and I find the "following defects :—

"The fall from the place where the soil-pipe enters the pot "drains is very defective, the level being higher in the centre "than at the termination.

"The drain-pipes themselves (six-inch pot drains) were full "of thick sewage matter, and had no luting or cement "between them. Lastly, at the very spot where the soil-pipe "is ccnnected with the pot drain pipes, there is a broken and "defective 'pot.'"

The w.c. was at the back of the house, and the drain ran under a cellar kitchen, not, as in the drawing, immediately under the hall floor.

PLATE XLIX.

Drain under house, with fall the wrong way. Broken pipe at the junction with the "soil-pipe."

PLATE L.

Disconnected and misconnected.

Mr. A. B., Town Clerk of the town of C., tells me that his house, situated 450 yards from the high road, was originally drained by nine-inch pipes into a pond a little beyond the high road. Early in 1876 the district was sewered, and the drain was cut off from the pond and connected with the main-sewer. In July, 1876, a maid and servant lad were seized with typhoid fever; the maid died and the lad recovered.

After the death, the drains were examined, and it was found *(a)* that waste-pipes from the kitchen, washhouse, pantry, and a lavatory, passed untrapped into the drains, with the illusory protection of a bell trap. *(b)* That the connection of the drain with the new sewer was so defective that the drain was blocked up at the junction, a nine-inch pipe having been inserted into an eighteen-inch pipe without any proper junction.

PLATE L

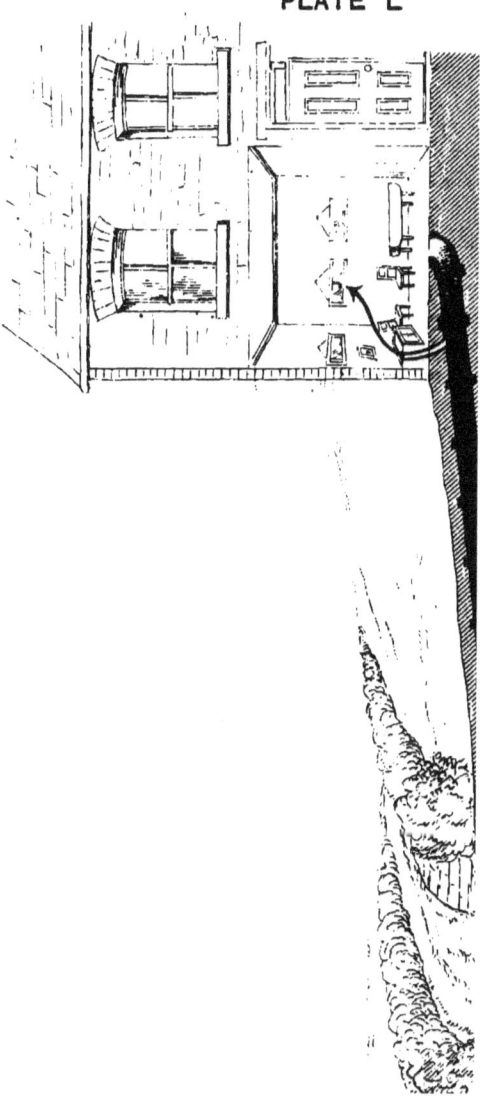

PLATE LI.

" To be continued in our next."

This example was also contributed to me by Dr. Britton, of Halifax, in the following note.

"In a gentleman's house the children were always ailing, "and in consequence I ordered an inspection of the soil-pipe "which was supposed to run under the house and some "outbuildings, and to join a main-drain in the road behind. "On the floor of the cellar and coal cellar being taken up, "there was found a very large quantity of sewage, which had "been accumulating *ever since the house had been built, seven* "*years before.*

"During the whole of this time all the sewage from the "w.c. had run under the floor of these cellars; for at the end "of the coal cellar the soil-pipe came to an abrupt conclusion "against a mass of solid rock, twelve yards thick, at the other "side of which a pipe was placed and connected with the "main-drain in the road. No doubt it was in order to save "the expense of blasting through the rock that the contractor "had scamped the work."

" The authorities saw the junction."

Until recently in Leeds, and probably in many a town besides, the following was the practice as to the inspection of sewers by the local authority. The Borough Inspector having received due notice from a builder of his intention to connect a house drain with a public sewer, came and "*saw the last pipe put in;*" with what security to the public may be judged from this Plate.

Nay, more, a builder from a neighbouring town told me that by a judicious tip he could dispense with even this formality, if it were inconvenient to suit the time of the Inspector.

PLATE LI.

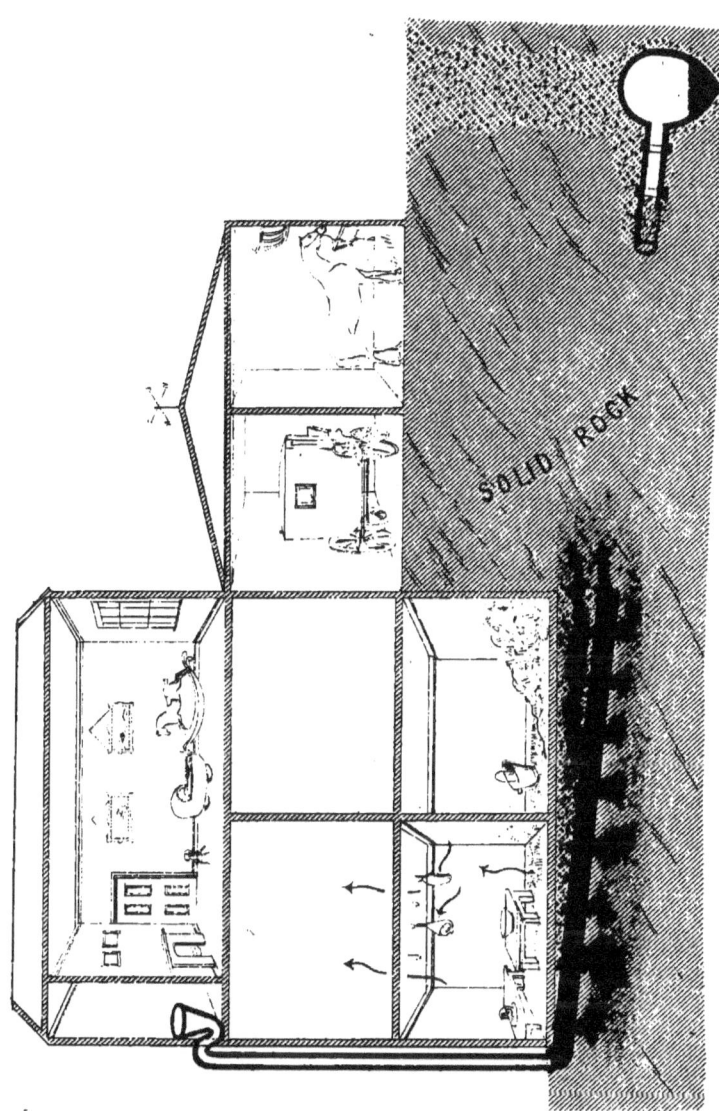

PLATE LII.

(A) Drain making the best of a rock.

The w.c. drain (A) is blocked as far as a rise in the drain, which was carried by curved tubes over the rock in order to avoid the trouble and expense of cutting through the rock. The fact expressed by this drawing, which looks like a caricature, was related to me by the landlord for whom the houses were built. Several builders who have seen the picture, have told me that they have seen drains so (mis) laid, and I know of one house in which this has been discovered to be the cause of obstructed drains since the publication of my lecture.

Since the publication of this plate I have been told, on many occasions by *eyewitnesses*, of the not unfrequent occurrence of this piece of rascality.

(B) w.c. discharging into the basement of a house.

The soil-pipe (B) missing the drain-pipe (C) had discharged the whole of the sewage into a triangular space below the ground floor. This went on for several months before the discovery of the defect was made, during which time "they never had the doctor out of the house."

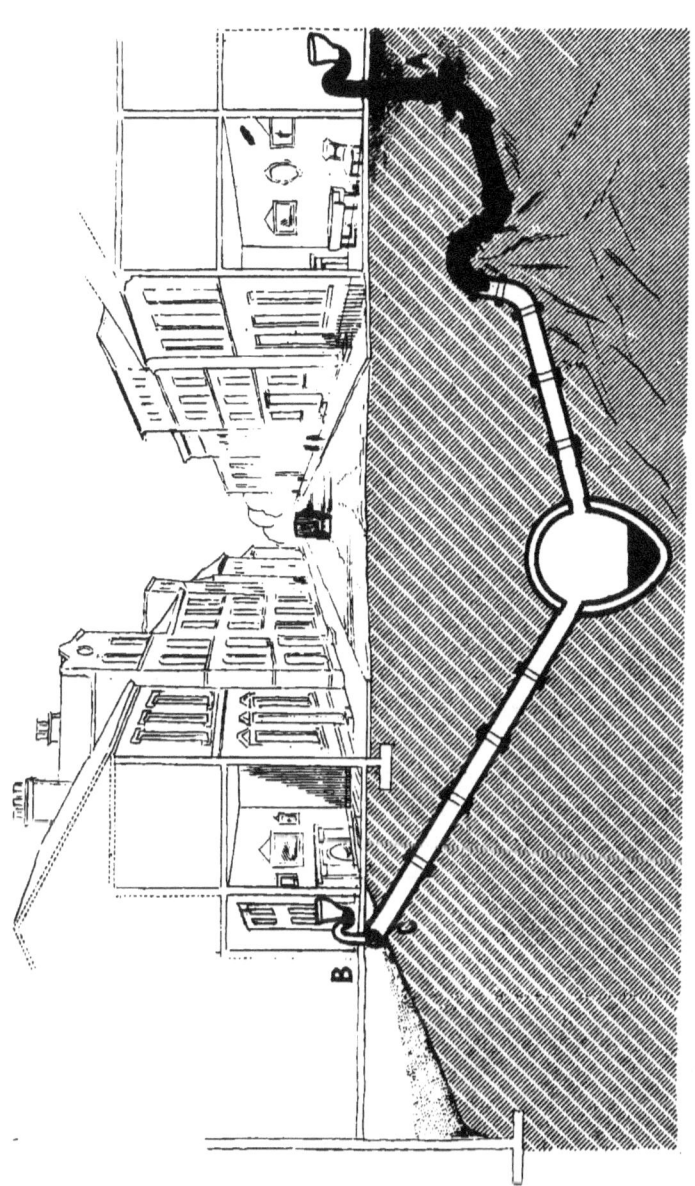

A. Drain "taking" a rock; sewage "refusing." B w.c. discharging into basement of a house.

PLATE LIII.

Economy in digging at the expense of "fall" in a drain.

This fact was related to me by a house agent and rent collector. A careless builder sometimes puts in the junction with a drain soon after commencing to build a house. When the time comes to lay the drain he finds that he has allowed far too little "fall." His duty would be to relay the drain and connection with the sewer with a proper incline. But this would cost money in excavation; so he saves his pocket, and leaves the drain to tell its own tale, when in due time the unlucky tenant finds his drains stopped, his house foul, his family ill, and the "tale told."

PLATE LIII.

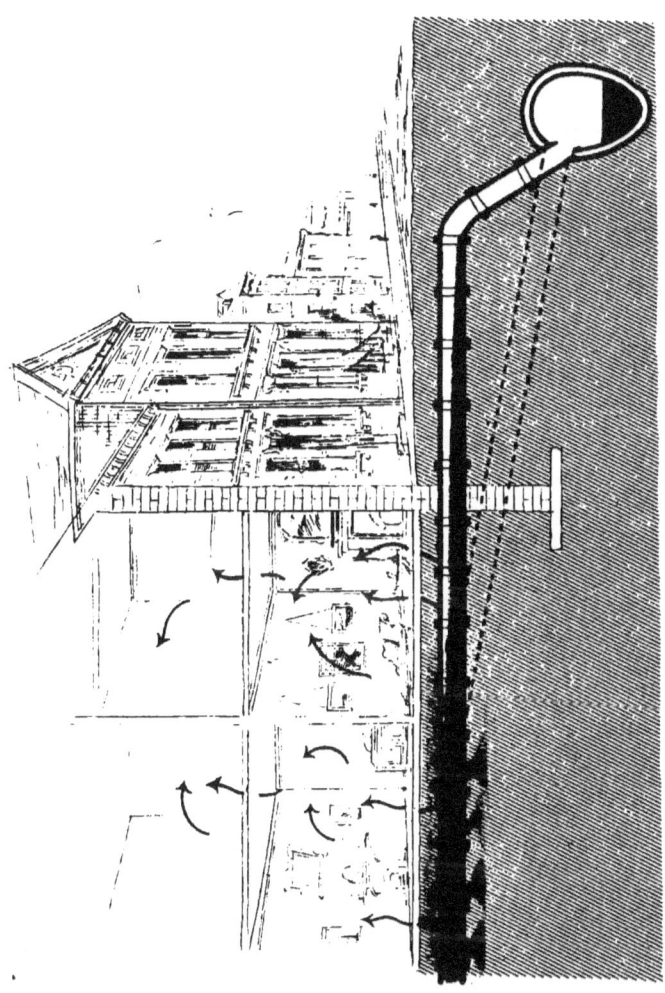

Economy in digging at the expense of fall.

PLATE LIV.

Six-inch pipe interpolated between two four-inch pipes.

This was discovered in some property which I bought six years ago. A cellar of one house was flooded by the overflow of an ash-pit, the drain of which was blocked up. The drain was followed, and traced to a junction with the w.c. drain of the next house. On enquiry, I found that this w.c. had long acted imperfectly, and no wonder, as the drain was blocked up for six feet, owing to the interpolation of a 6-inch between two 4-inch pipes.

A gross instance of this mode of scamping is thus described to me by Dr. Murray, of Burley-in-Wharfedale :—

"Some villas were drained by 12-inch pipes, which passed "along the road into a field where a cesspool *ought* to have "been constructed (according to plans passed by the Local "Board). Having carried the 12-inch drain into the field, "the contractor, coming across ordinary 2-inch draining "tiles, connects the larger with the smaller pipes, and fills up "the trench. In a short time a block takes place, and the "sewage bursts up into the field. Two cases of typhoid "fever occurred in one of the houses."

PLATE LIV.

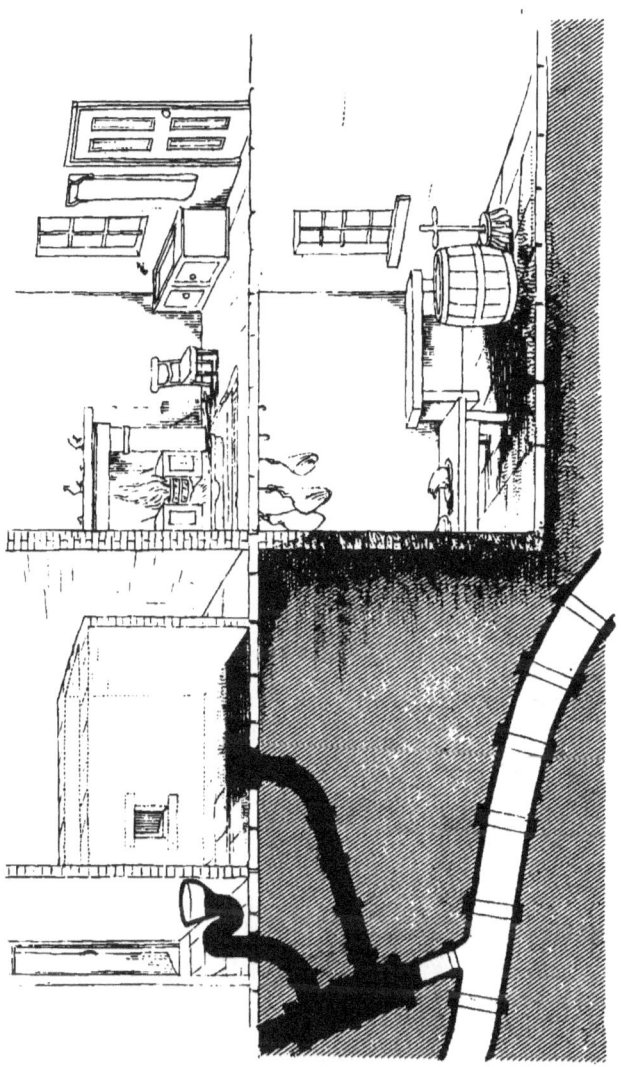

Six-inch pipe between two four-inch pipes.

PLATE LV.

Road scrapings and ash-pit refuse for mortar and plaster.

This picture represents what has been, I fear, only too common an occurrence of late years in Leeds. Road scrapings from our Corporation depôts, and the emptyings of common ash-pits instead of loads of clean mill cinders, have been ground up along with a bare pretence of lime, to make the mortar for setting the bricks, and the plaster for covering the walls of miserable tenements.

This mud-made mortar sets so slowly, that the builder has to prop the wall (this I have seen), and, as I have often been told, has to light fires against the wall to " encourage " the mortar to set.

Walls plastered with such rubbish are slow in drying, have large greasy patches which strike through whitewash, crumble when a nail is driven into them, and probably are a prolific source of the illnesses from which people suffer who inhabit newly-finished houses.

" If you bray a nail into the wall, half of it comes down "—said a Leeds victim, suffering from disease of lung, probably brought on by the unwholesomeness of the walls of his house.

The following fact was told by a leading Leeds builder to the gentleman who related it to me :—

" In about 60 new 'speculators'' houses not a single load " of clean lime was used—mortar and plaster were made of " lime which had done duty in tan pits "—therefore spent, and full of animal cleansings. The builders of the houses were also the owners.

PLATE LV.

112

"Road muck" and "midden refuse" for mortar and plaster.

PLATE LVI.

Terrace of the Future on the Refuse of the Past.

This plate needs but few words. Until recently, no check has been put upon the haste of speculating builders, who have built thousands of houses on unhealthy rubbish heaps, long before the animal and vegetable refuse has had time to ferment, decompose, and cease to be poisonous. Within the last few years, a plot of land, which served as the depôt for the road scrapings of the Corporation of Leeds, has been covered with houses and shops. Such proceedings will surely be impossible in the future; thanks to the new Building Bye-Laws of our town. *Vide Appendix*, (§ 4).

For a vivid description of "foul made-ground," let anyone read in Scribner's Magazine for May, 1881, an article " On the Sanitary Condition of New York."

PLATE LVI.

"Terrace of the Future on the Refuse of the Past."

PLATE LVII.

Hunting for drains.—No plans.

This plate is intended to enforce a *lesson* and to proclaim a *fact*. The *fact* is, that it is extremely rare for the owner of a house, still more rare for the tenant, to possess a plan of his drains. A house is built, and sold, and occupied, and after the lapse of a few months, or it may be a few years, the drains are blocked, and need examination, and no clue can be found to their whereabouts. The architect, perhaps, is dead, the builder a bankrupt, and the workmen are dispersed. The *lesson* is that every house ought to have attached to it a plan of the drains as a matter of right and law.

PLATE LVII.

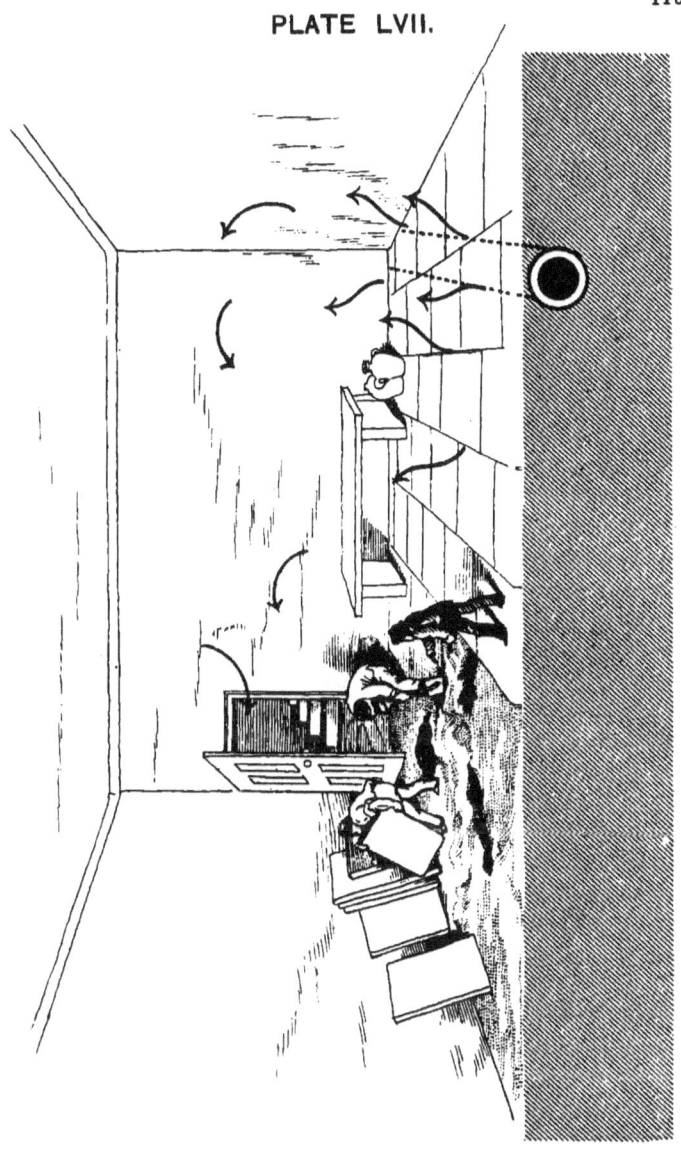

"On the wrong scent."—No plans of the drains.

PLATE LVIII.

Drains blocked by willow roots.

For the sketch and fact on which this drawing is based, I am indebted to the Rev. A. C. Black, of Burley-in-Wharfedale:—

The main sewer of Burley was found to be blocked close to the wall of the Vicarage garden. The roots of a very fine willow tree had penetrated the joints of the drain pipes, and having thriven and increased in the congenial feeding ground within the drain, had caused a block.

A second instance of a drain blocked by willow roots was communicated to me by Mrs. Priestley, of Hertford Street, Mayfair.

A third instance came to me from the Rev. Stephen Saxby, of East Clevedon, who once found 20-feet of drain pipes filled by willow roots.

Moral.—When laying drains in wet ground near to willow trees, unite the pipes by cement, and not by clay. The fine roots can penetrate the clay, and so gain access to the drain. As a rule, however, clay is preferred as luting, being constantly moist and not liable to crack.

PLATE LVIII.

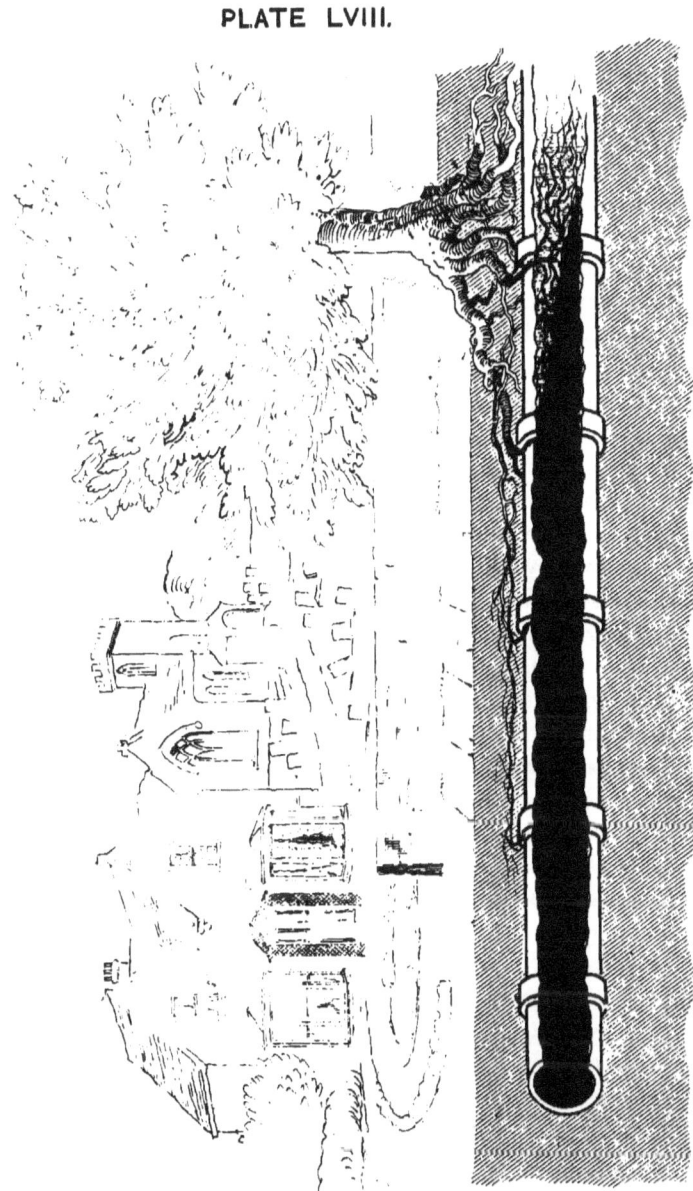

"The battle of the willow trees." Drains blocked by willow roots.

PLATE LIX.

Cesspools under London Houses.

A London physician contributed the sketch from which this drawing is made. On investigating the sanitary arrangements of his house, he found no less than five cesspools and their connecting drains under the basement of his house. The w. c. was in the centre of the house, joining one of the cesspools in the cellar. On shewing the sketch to another physician, neighbour to the friend who supplied this fact, he told me that he discovered the very same arrangement of cesspools in his own house.

A lady friend writing from London says :—" In many " parts of London the old cesspools still exist, and are only " discovered by falling in. (*Vide* Plate XXXVIII.). These " were never fairly done away with when the new system of " drainage was introduced. In Edinburgh numbers of the " houses still have cesspools, which are never emptied until a " blockage takes place, or sickness breaks out. My youthful " recollections of Edinburgh in my father's handsome house, " most beautifully situated, recall a basement which must have " been flooded with sewage. I now know that the drainage " of the house went into a cesspool which no one ever " enquired into, and which never was emptied during the " whole of my youth. We were always having fevers, but it " was accepted as a thing natural to youth. These reflections " are dreadful.

" When a house in Mayfair was being done up about " four years ago, the men were digging at the foundations, " when they suddenly broke into a horrible pit, the effluvia " from which sent them flying in all directions. It was then " found to have been a pit into which cattle had been thrown " during a murrain when Mayfair was a large farm."

PLATE LIX

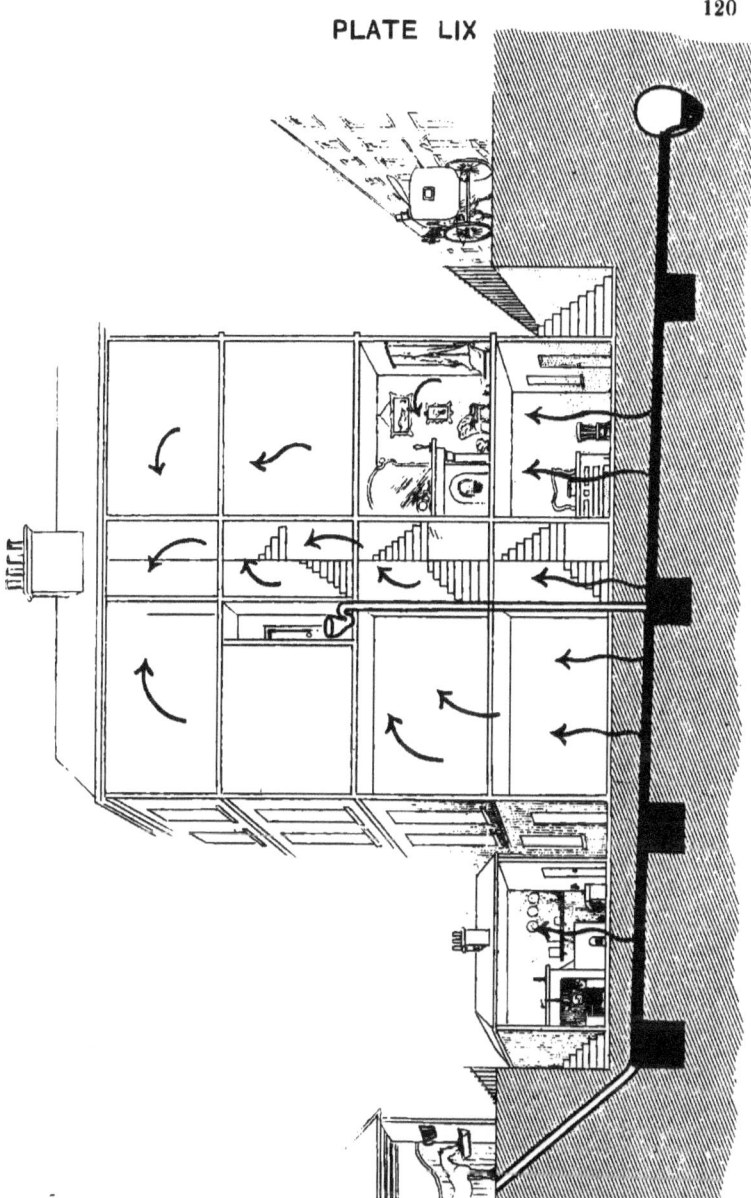

Not a hundred miles from Harley Street.
Five cesspools under a London doctor's house.

PLATE LX.

Vicarage rendered unhealthy by adjoining graveyard.

For this fact I am indebted to the Rev. A. C. Downer, Vicar of Ilkley. It was communicated to him in a letter as follows: —"I strove to serve three churches without even one curate, and when I was already nearly broken down with this, infiltration from the churchyard into our cellars caused much and grievous illness amongst us; and the mischief proving incurable, we were finally driven out of our lovely vicarage altogether. I had for months been prostrate with low fever and ague."

PLATE LX.

Vicarage rendered unhealthy by infiltration from churchyard.

PLATE LXI.

Cellar damp from slop water.

In a vicarage in a Yorkshire country town the cellars were constantly "standing in water," and were so damp that they could not be used. The water was supposed to be surface water, and the question of an expensive drain to intercept the water was considered.

The Vicar's child, being in ill health, was taken to a watering place and put under medical treatment. The doctor suspected insanitary conditions at home, and at his suggestion an efficient sanitary inspection was made, and the conditions represented in the plate were disclosed.

The housemaid's sink passed down the wall of the kitchen to join a drain, running between the arched ceiling of the cellar and the kitchen floor, where the drain passed over one of the piers of the cellar; the drain pipes had settled down and opened at the joints, thereby discharging all the slops from the housemaid's sink through the middle of the pier into the foundations of the cellar floors

PLATE LXI

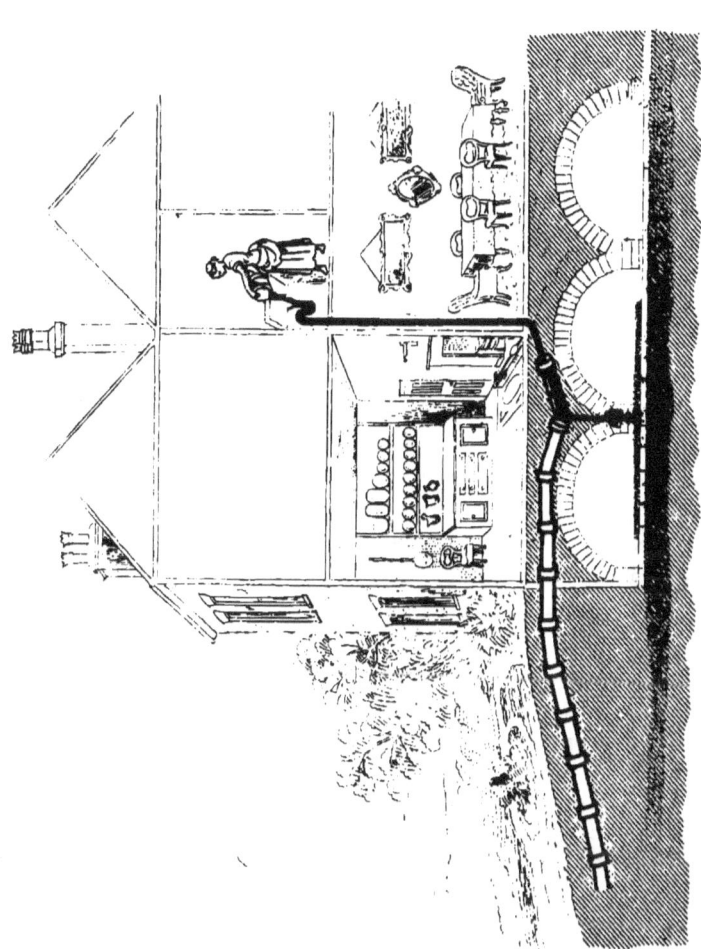

Cellar kept damp twelve years by slop water.

PLATE LXII.

A villa at Cannes.

Not many years ago a lady was advised to winter in the South of France for the improvement of her health. Whilst she was residing in a villa at Cannes, her maid fell ill of typhoid fever. The lady at once took a sanitary survey of the house, and found that under the maid's room there was a foul w.c., which discharged its contents into a large tank in the room below. The overflow from the tank was soaking the floor of this room, which was next to the larder and close by the kitchen.

Moral.—"Coelum non animum mutant qui trans mare currunt." "Dangers to health" are not lessened by going to Continental health resorts.

The more strongly English public opinion realises this fact and makes its dissatisfaction therewith felt, the sooner will the authorities of these Continental cities feel that in self-interest they must study and set right the grave sanitary deficiencies which from time to time are but too painfully forced upon the notice of the English public.

PLATE LXII.

A Villa at Cannes.

PLATE LXIII.

A Scotch mansion let for the season.

The fact expressed in this picture was told to me by a colleague. He was called into Scotland to see a lady ill in puerperal fever, her husband having taken the house for the double purpose of a summer holiday and also of securing a healthy place for the event which was expected.

The kind of illness suggested insanitary conditions and an investigation. The result of the inspection of the house was communicated to me by the father-in-law of the lady as follows.—" There were nine w.c.'s., six of them in the centre of the house. All the communications from them were under the house and passed to the front, where the main drain went. About forty yards from the house the pipe from the laundry joined the main drain, and here a stoppage was found. Beyond this junction the drain was continued to the river three quarters of a mile. I examined the drain beyond the stoppage, and was certain that no sewage had passed for a long time. I was satisfied that the pipe near the house was only a cesspool."

Moral.—In selecting a house for a summer holiday don't forget to look after the drains, and if possible have the house examined by an inspector on whom you can rely.

PLATE LXIII.

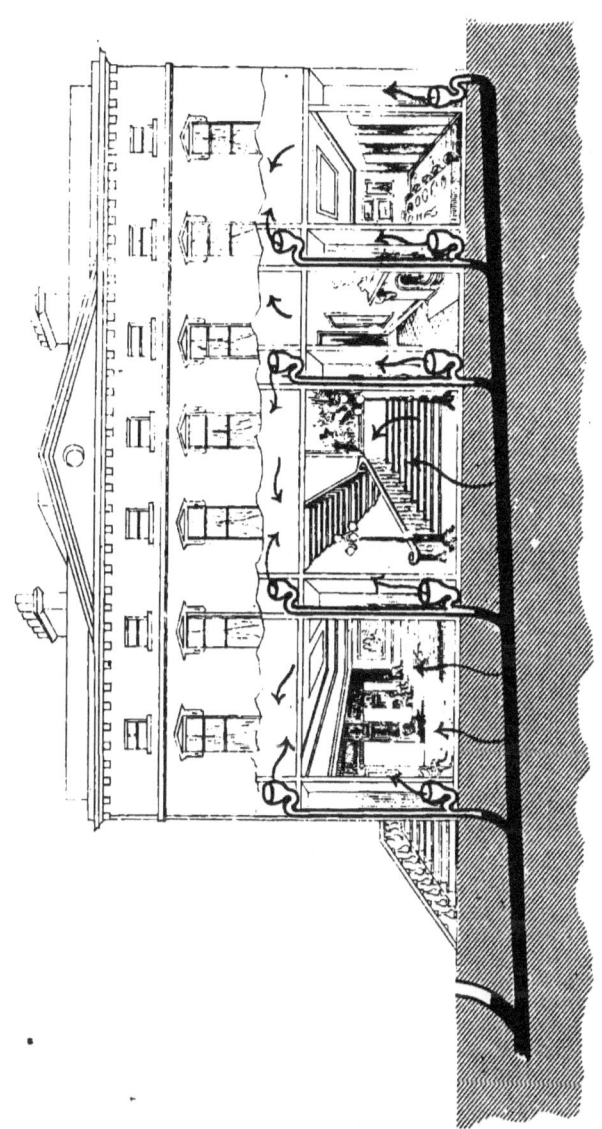

An "Eligible Mansion" in Scotland let for the season.

PLATE LXIV.
Shooting Box in the Highlands.

The wife of a medical friend of mine preceded her husband to the Highlands that she might set in order the shooting lodge before the arrival of her husband and his guests. Being well informed in sanitary requirements and deficiencies, she did *not* take it for granted that her orders had been strictly carried out, that the drainage should be put into perfect order.

That she might test the main drain, she posted herself at the empty cesspool, whilst her daughter, stationed at the house, gave a signal, when several buckets of water were poured into the drain; after long waiting some driblets of water reached the cesspool, not through the drain, but through chinks in the side of the cesspool.

Inference. Drains must be blocked.

Workmen came, found the drain running up-hill and blocked by roots of willow tree. (1. A.)

The drain was relaid with a proper fall, (1. B.) The pipes were jointed in cement instead of clay in order to shut out willow roots and such like intruders, and five openings to the surface were provided in order to secure ample ventilation of the pipes. The lady meantime sat by knitting and watching the whole of the work.

Figure 2 is drawn, in error, too much like the other two. It is intended, however, to represent a separate drain from the laundry to a separate cesspool. A fortnight after the assembling of the visitors, a stench was complained of by the laundry maids. On investigation it was found that the pipe by which the laundry cesspool overflowed was at a higher level than the pipe through which the cesspool was filled, the result being that before any overflow could take place the level of fluid in the cesspool must rise above the inlet pipe and close it to all escape of air and sewer gas.

For an interesting history of this "shooting box" and its sanitary renaissance, read "Our Highland Home," published by the National Health Society, 44, Berners Street, London. Price 1d.

PLATE LXIV.

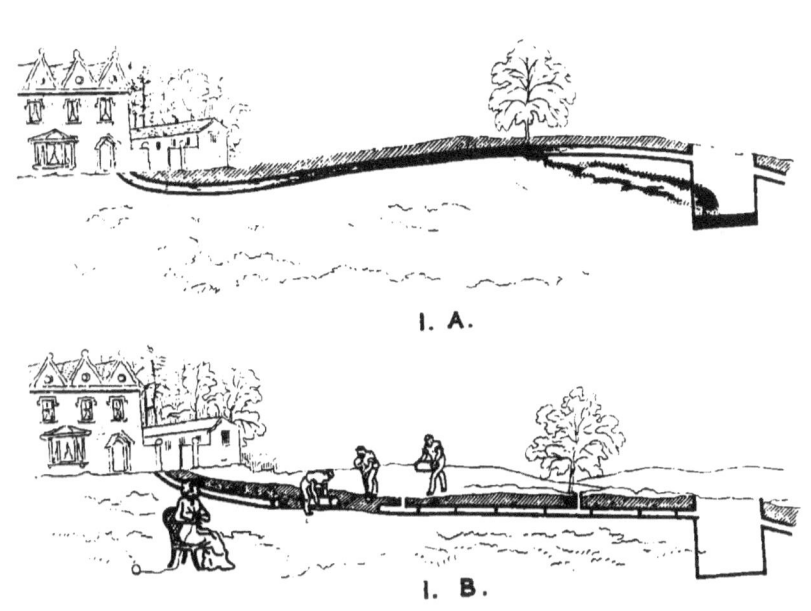

I. A.

I. B.

" One eye for her work, and another for the workmen.

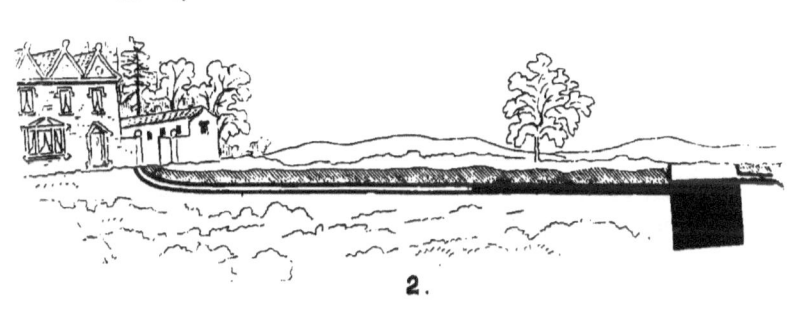

2.

A Highland Shooting Lodge.   K

PLATE LXV.

Manure heap piled against the wall of a house.

Although I have no authentic facts to relate in proof of the dangers involved in the conditions here depicted, I have strong reason to believe that in two instances they were the source of diphtheria.

In a farm house, high up on the moors of Yorkshire, a case of diphtheria occurred. There was no house near it, and no known source of contagion.

In another instance in a small farm house high up in the Welsh hills, diphtheria broke out.

The only obvious insanitary condition (and this common to both) was the condition of the stable yard, the manure being habitually piled up against the wall of the house, so that both wall and floor were damp.

Such conditions are unsafe, even though they may not have been, as I have surmised, the cause of the diphtheria.

A medical friend told me of illness in a house which he attributed to a similar cause. All the sanitary arrangements of the house had been carried out with great care, the w.c.'s, sinks, and baths being disconnected, but the side wall of the house and of a cellar beneath had recently become damp, from a large heap of manure placed against the side of the house by the gardener, and used as a hot bed.

PLATE LXV.

Manure heaps against house walls.

PLATE LXVI.

Vaccination and drainage faults.

When a case of death or serious illness occurs after vaccination, it is generally seized hold of by "misguided and imperfectly informed persons," who make capital out of it, in order to induce people to believe that vaccination is injurious and useless. In the trials that such persons get up the fact is often brought out that several infants have been vaccinated from one source, and that, whilst one infant has suffered seriously after the vaccination, the rest have passed through it without a drawback. It is clear that, in the one exceptional case, some factor distinct from the vaccine is needed to explain the result.

The following facts, related to me by Mr. Edward Atkinson, of Leeds, may throw light on such cases.

A healthy child, aged 4 months, went on well until the 9th day after vaccination, when it became feverish, and abscesses formed in the finger and ankle. The illness suggested an inspection of the drains, when it was found that the waste pipes of a lavatory and bath near the nursery were untrapped, and passed direct into a soil pipe.

Dr. Britton, of Halifax, tells me of an instance in which erysipelas attending vaccination was traced to an open cesspool just under the nursery window.

PLATE LXVI.

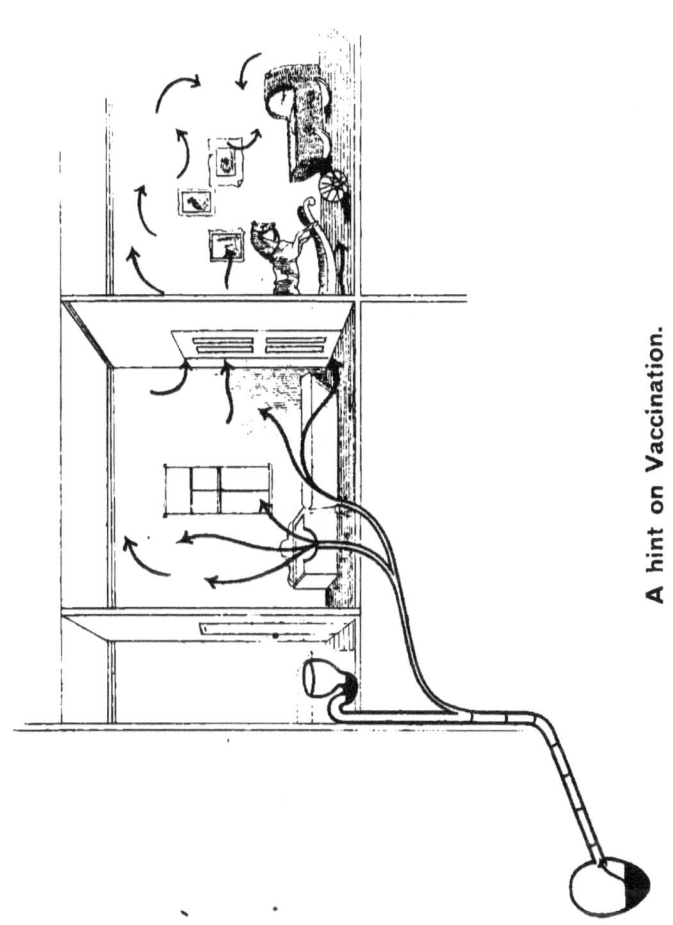

A hint on Vaccination.

PLATE LXVII.

Arsenical Wall Papers.

This danger cannot well be expressed in a drawing. In order, however, to keep to the fundamental principle of the book, viz., to appeal to the eye in order to enforce every lesson, this plate is given as expressing the fact of arsenical paper being stripped off a wall.

Much has been written in medical and lay journals of the injury to health inflicted by arsenic in wall papers.

During the last four years I have traced ailments to this cause in several instances, and I keep, as trophies, pieces of the detected and condemned papers.

About seven years ago my own children were unwell from sleeping in a newly-papered bedroom. The paper had a brilliant green pattern, and was guaranteed "free from arsenic." The illness of one child after another led me to have the paper examined by my friend Mr. Scattergood, and he reported the paper full of arsenic in a loose and dangerous form. The paperhanger was dismayed, replaced the paper, and, I believe, no longer takes "warranted" papers on trust.

Since the publication of the first edition of this book, the dangers from arsenic in wall papers and other articles of every-day use have received ample illustration from the facts most industriously collected by Mr. Henry Carr, 21, Cedars Road, Clapham, who has written a pamphlet and lectured before the Society of Arts on the subject. Speaking of this illustration, he says "you want another figure. A man going away—can stand it no longer. This is a common fact."

Akin to this subject is the filthy custom of placing a new paper on a wall without stripping off the old one. In one instance five papers were removed from the wall of a room, the occupants of which had been constantly ill.

PLATE LXVII.

Stripping off Poisonous Wall Papers.

PLATE LXVIII.

Admission of Fresh Air and Exclusion of Dirt.

It is with some diffidence that I venture to offer remarks on ventilation.

Without entering on a discussion of the merits of various plans proposed for admitting fresh air into rooms, I will state what has been done in a house, specially fitted for the use of invalids, and in my own consulting-room. Bearing in mind the teaching of Plate III., that the chimney has to be supplied with air, a Tobin's tube, with a **sectional area about equal to the chimney pot**, was placed in each room. The effect of this is that the rooms are constantly fresh night and day, that irregular draughts are much reduced, and that, except in cold weather (an outside temperature below $32°$), the ventilators are rarely closed.

Having secured for each room its own supply of air for the chimney, the next question was, how to **clean the air, and exclude the dirt**. I had long seen that, if air is to pass through a screen without retardation of the current entering the room through the tube, the area of the screen must be many times (perhaps 10 or 15 times) the area of the section of the tube. Acting upon a suggestion of Messrs. Bapty, of Leeds and Bombay, I requested Messrs. Harding, of East Parade, Leeds, to place a screen, if possible, in the tube itself, telling them that the screen must be at least ten times the area of the section of the tube, and that the section of the

PLATE LXVIII.

Ventilation without Dirt.

tube must equal the section of the chimney pot. Mr. Joseph Harding very shortly hit upon the happy idea of placing the screen (B) in the tube diagonally from top to bottom, and thus achieved what I was seeking.

Recently, Messrs. Harding have invented a means of admitting air into a room without draught, named a "Diffuser" (E). It is a contrivance by which the fresh air is shot into the room through a series of short tubes placed in the front and sides of a box. This box being placed near the ceiling, the cold air mixes with the warm air, and thereby no draught is felt. The form of ventilation, therefore, which I have found to answer best is a combination of Harding's "Diffuser," with the broad flat tube containing a screen. The arrangement by which the whole front of the tube opens on hinges (*vide* door C), is an improvement made by Messrs. Harding since the publication of my third Edition. It has two advantages: the first, that the screen can be more readily removed for cleaning; the second, a very valuable point, that in summer, when there is no fire in the room to overcome friction and draw a good current of air through the screen and "Diffuser," this door can be left wide open, when, under the pressure of air outside the building, there is an almost constant current through the screen. It is virtually a window open day and night with "no admittance to dirt." To render the exclusion of dirt more perfect, the window is shut down upon a strip of carpet or plush, and the junction of the two window sashes is pasted up with paper. The result is that in the centre of a dirty, smoky town I can keep my library free from smits. I am satisfied that by means of this apparatus we can secure in a town **freshness of atmosphere, absence of draught,** and **exclusion of dirt.**

(A) is the grate in the outer wall, to keep out birds and mice. This grate must not "throttle" the air, *i.e.*, must not admit less air than the tube it has to supply can carry.

(B) is the screen covered with canvas or bunting. It slides in grooves, and is removed twice a week that it may be brushed by a soft brush, or the meshes would be choked.

(C) is a door to allow the screen to be withdrawn for the purpose of cleansing.

(E) Harding's "Diffuser."

Harding's "Diffuser" is patented, but the screen is not patented.

Floral Art Ventilator.

The Floral Art Ventilator is an elegant contrivance for introducing fresh air into a room by the open window without draught. An inner casement serves the double purpose of a screen, and a conservatory, within which growing plants purify and moisten the air, as it passes upwards.

It has been designed by the wife of a physician living in Mayfair, and is most artistic in effect. In winter it may act as a double window to keep out the cold, or as a fire-screen to keep off the heat. The invention has been registered by the National Health Society, to whom it was presented, and has been introduced to the public by Messrs. Howell & James, of Regent Street, who are sole licencees.

Finally, let me remind my readers that all passive ventilators for the admission of air depend essentially for their efficiency upon the indraught of fires in a room or house, or upon wind pressure outside the house.

PLATE LXIX.

Why Glass Cases don't exclude Dust, and how to make them do so.

Dust is the ruin of collections in museums, and a perpetual source of most annoying expense. It is a discredit to science that we have not conquered such an extravagant enemy, and yet I feel sure that the remedy is a simple one, if we will but ask ourselves the question: why does dust always enter the most carefully made glass cases?

The answer is clear. The air inside the case is constantly altering in volume, under changes of temperature, and changes of barometric pressure. This perpetual variation causes the entrance of perpetual currents of dirt-laden air through minute crevices. What, then, should be done? First and foremost, the fact must be acknowledged, and a sufficient air channel made, so that (as in Plate III.) the air may enter by the "legitimate" channel, and the "irregular" channels may cease to act; next, the "legitimate" channel must screen the air. For achieving this let me venture to make several suggestions.

Suggestion A.—This is an inverted square tube, of a section 4 inches square, attached to the side of a glass case in the Museum of the Leeds Philosophical and Literary Society. The mouth of the tube is filled with *lightly*-packed cotton wool. In a few months the outer part of the wool was blackened with dust. Such a tube, however, is probably far too small.

Suggestion B.—That one or both ends of a glass case be closed with wire, for safety, and the wire covered with baize or bunting, which would admit the air and exclude the dirt.

Suggestion C.—This, if it would act, would be the most scientific, most self-acting, and most perfect. It is based on a suggestion of Dr. Eddison, of Leeds:—Having ascertained from Professor Rücker that the volume of air in a case will vary in volume about one-tenth, it occurred to me that the back of a glass case might be made double, the distance between the two backs being equal to one-tenth of the depth of the case from front to back. Then if the "inner back" stopped short of the top of the case by two inches, and the "outer back" stopped short of the floor by two inches, there would be free ingress and egress of air between the interior of the case and the space within the double back, but the outer dirt-carrying air would never directly reach the interior.

PLATE LXIX.

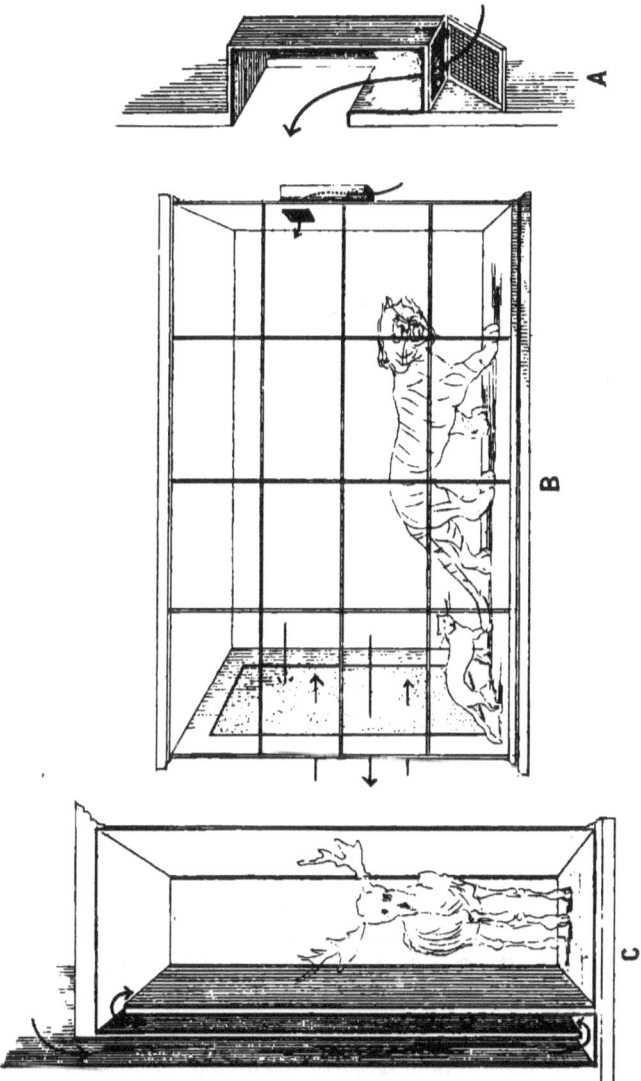

Dust the ruin of Museums. Why not keep it out?

PLATE LXX.

"Window Ventilator" in the Roof of a Brougham.

Having, during the last eight or nine years, derived much comfort from the window ventilator, I wish to publish this for the benefit, more especially, of my medical brethren. As many of them spend a great part of their life inside a carriage, it is for them highly important :—Firstly, that they should breathe as pure an air as possible, and that without the infliction of a draught :—Secondly, that they should be able to read with the best light attainable, a roof light, and avoid the distressing variations of the side light in passing through the streets.

The idea of a roof light was suggested to me by my friend Mr. R. P. Oglesby. On giving instructions for the roof light to the carriage builder, Mr. Bradley, of Leeds, he suggested that the window should be on a hinge, and should open backwards, and thus supply ventilation. The result exceeded my expectations. The following points must be attended to in its construction :—

The size of the window should be about 18 inches by 8.

The position should be vertically over the place in which the book is held, *i.e.*, over the knees. This is important for three reasons—the first, that it is the best position for illuminating the book; the second, that if there should be a strong wind from the rear, no draught can come upon the head; the third, that if during rain an occasional drop of water enters, it will not fall on the cushion, but on the floor, or on a rug on the knee.

PLATE LXX.

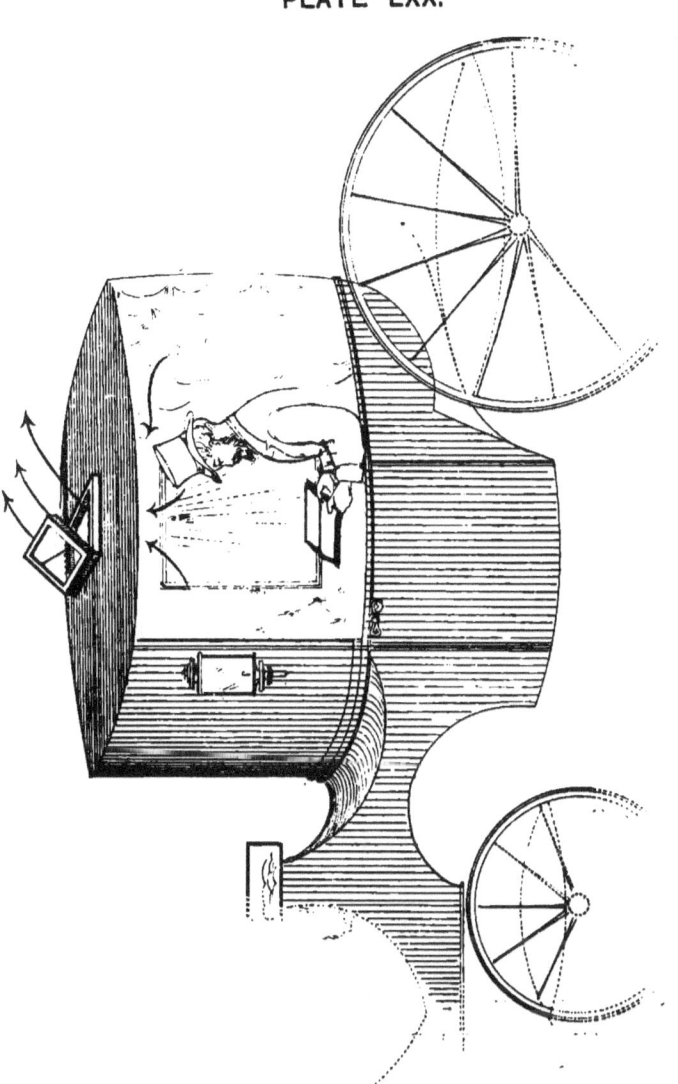

Window Ventilator in the Roof of a Brougham.

The elevation is secured by a small rack and prop.

The closure, (very rarely needed,) is important. If the window be fixed closely down, the vibration sucks in water during rain, and causes dripping. This is avoided if a hook fastener fixes it one-sixth of an inch open.

In winter the air of the brougham remains quite fresh, even with three persons, without the need of opening any side windows. The following experiment is interesting:—Travelling one frosty day with two companions, and observing the windows of other carriages dull with "steam," whilst my own were clear, I closed the roof ventilator, and in *five minutes* the whole of the windows were covered with steam. The ventilator was then opened, and in five minutes more three-fourths of the windows were clear.

May not much of the delicacy of hard-worked medical men be caused by their breathing in their carriages a deteriorated air, with the alternative of draughts, which their enfeebled health can ill endure? May not such a ventilator enable them to throw away the respirator so commonly used?

Several medical men in Leeds and elsewhere have adopted the roof ventilator.

I recently (1883) discovered that, in the absence of a strong wind to the rear, and when the brougham is in motion, the current of air is entirely outward at the ventilator, thus providing a more rapid change of the air than I was aware of, the ventilator acting as an "extractor." In fact, when the carriage is moving, the air enters by the imperceptible crevices around the door and windows from pressure of the air outside, and escapes in a full current, as is shewn by a lighted match, by the ventilator, thus rapidly changing the air without a perceptible draught.

ADDITIONAL DEFECTS
NOT ILLUSTRATED.

I.
W.C. ventilated into false roof.

(Communicated by Mr. A. W. M. Robson, of Leeds).

Mr. A.'s children had good health until they came to live in Leeds, after which they were constantly ailing, and one child died of infantile remittent fever. The room containing the w.c. was ventilated by a shaft into the false roof, beneath which was the nursery. After this shaft had been carried completely through the roof into the open air, no further sickness occurred during the remainder of their stay in Leeds, which was about two years.

II.
Soil-pipe ventilated into false roof.

(Communicated by Mr. Dale James, of Sheffield).

A gutter in the false roof originally conducted the rain-water to a cistern inside the house. When afterwards the cistern was fed from the public water supply, the rain-water was diverted from the cistern in the following manner:— Instead of being conveyed by a fall-pipe outside the house, it was allowed to escape by an open-mouthed pipe ending in the top of the soil-pipe, thereby allowing the ascent of all the sewer gas from the soil-pipe into the false roof, whence it was drawn into the house.

A second instance of the kind is related to me by Mr. M. M. McHardy, of Savile Row, W. In this case an open gutter in a bedroom concealed only by boards, after receiving the overflow of a cistern and the slops of a sink, discharged into a rainfall pipe, which, after receiving the soil-pipe of the w.c., ended in a closed drain. The drain and soil-pipe were thus put into direct open communication with the upper rooms of the house.

L

III.
Defective junction of ventilating pipe with soil pipe

Not long ago an eminent lawyer and his wife died of typhoid fever.

After this event, it was discovered that a ventilator was taken from the upper bend of the soil pipe inside the house, and that at the point of junction the soldering was defective, so that the soil pipe really was ventilated directly into the house.

IV.
Cesspool directly below bedroom window.

(Communicated by the late Dr. Moore, of Lancaster.)

A young gentleman who was in the habit of sleeping with his window open, was always ill when he came home and occupied a certain room. After a time he changed his room and the malady ceased. The discovery was then made of a large cesspool immediately below his former bedroom window.

V.
Foul smell in drawing-room introduced by air-brick under floor.

(Communicated by Mr. Henry Carr, of 21, Cedars Road, Clapham, who has kindly sent me a sketch.)

Foul smell in drawing room. Hollow space under floor ventilated by "air bricks." Opposite an "air brick" an imperfect joint of a fall pipe. Fall pipe leading into a drain and cesspool.—"My attention was first drawn to the rain water pipe by observing a cobalt blue mark just above the joint." In this case it would seem that the sewer gas, as it escaped from the joint of the fall pipe, was drawn through the air brick under the floor, and then between the boards into the room.—(*Vide* Plate III.)

VI.

Air-space under dining-room floor, used for rubbish.

(Communicated by Mr. R. N. Hartley.)

Air-space left under dining room with through ventilation to keep foundations dry. From this a trap door opening into wash kitchen. The former occupants of the house seem to have made this space a convenient receptacle for all kinds of household rubbish of various degrees of offensiveness.

VII.

A Continental Hotel.

(Communicated by Mr. Joshua Hartley, Surgeon, of Malton.)

A hotel abroad with central court (roofed in). In the middle of the floor an open grating leading untrapped into a drain. Many bedroom windows opening into the court.

VIII.

Drain blocked by old wall papers.

(Communicated by Dr. Fitzgerald, of 8, Palace Road, S.E.)

An instance of illness, produced by the obstruction of a drain in a newly painted and papered house. The workmen had disposed of the paper which they scraped from the walls by putting it into the drain.

IX.
House drains sealed up during the making of a new sewer.

(Communicated by Dr. Clifford Allbutt)

Cases of drain illness in a house ; the doctor in attendance blames the drains; urgent denial ; re-assertion ; investigation. It was found that, in making a new main sewer below the level of the old sewer, the workmen had sealed up the portion of the old sewer which received the drains of the house, without making any attempt to connect these house drains with the new main sewer.

X.
Neglect to connect new drains with sewer.

(Communicated by Mr. Robert Hagyard.)

A new and carefully planned system of sewers was laid down in a small town not far from Leeds. The soil-pipes of two w.c.'s, together with the sink waste-pipes from a row of six houses, were united into one drain which was carried to the main sewer. For eight months the sewage collected in these pipes, and eventually burst into the cellar of one of the houses, which was found to be filled to a depth of two and a half feet with nearly solid sewage. It was then discovered that the drain of eight-inch pipes had been conveyed to a six-inch junction in the main sewer, and that the disc closing, temporarily. the junction had never been removed.

XI.
Foul smell drawn into kitchen through disconnected drain pipe.

(Communicated by Mr. Wm. Wailes).

Bad drain smell in kitchen, especially when a large fire was used in cooking. Untrapped waste of sink delivered into the open air over a grating which led untrapped into a sewer, and, under the influence of the strong indraught of the fire, conducted the foul smell into the kitchen.

XII.
Soil pipe within a house laid open by a falling brick.

(Communicated by Dr. Swanwick, of West Hartlepool).

In a house in which great care had been taken to disconnect waste and sink pipes, first one daughter, and then a second was taken ill with sore throat. The drains, though considered perfect, were examined, and a large hole was found in the leaden soil pipe, against which was seen a fallen brick. The soil pipe, of course, had been boxed off and concealed.

XIII.
W.C. opening out of bedroom. Erysipelas.

Nearly 25 years ago my father attended a gentleman who had received a compound fracture of the skull. The gentleman went on well for three weeks, and then had erysipelas and died. Recently I was able to enquire about the sanitary surroundings. A lady who helped to nurse the patient had to leave a few days before the gentleman's death, as she was ill with sore throat. She told me that a w.c., generally disused, and brought into use during the illness, opened directly into the bedroom.

I have been told of a large country house in which, in the state bedroom, there is in the middle of the room a w.c. disguised as an ottoman. Comment on the danger is unnecessary.

XIV.

Horizontal transmission of sewer gas to a room distant twenty-two feet.

(Communicated by Dr. Oliver, of Harrogate.)

In the winter of 1878 the atmosphere of his consulting room became unpleasant, especially at the end nearest the kitchen. Drains were suspected and overhauled, but no defect was discovered. There was no drain near the room. In the following spring matters became worse, and Dr. Oliver had a serious illness of the kind produced by drain poison. This led to the examination of the drain in the scullery, which was separated from the consulting room by the kitchen. The boards were on this occasion taken up, and a hole was discovered in the pipe passing from the sink to the drain. Previous investigators had been content to look at the trap under the sink and "above board," and had forgotten or never known the lesson taught in Plate XIX. This defect was remedied, and all unpleasantness in the consulting room ceased. It would seem that the sewer gas had been conveyed from the scullery, and past the kitchen, along the rafters supporting the floor to the corner of the consulting room, a distance of twenty-two feet, and had penetrated the room behind the skirting board, which had been loosened by hammering during the laying down of some hot water pipes.

XV.

Drain blocked by "vermin" traps or grates.

Two instances of this intensely stupid arrangement have been brought under my notice.

In order to prevent rats from running up a drain into the house, a grating is sometimes inserted into a drain, the ingenious authors of this wonderful mechanism forgetting that bars which shut out rats will also shut in the solid portions of sewage, which in time collect against the grating and stop it up completely.

The first instance occurred many years ago in Leeds, and was related to me by the son of the occupier of the house. His father had a severe attack of erysipelas, and recovered. Not long after he had a second attack, which proved fatal. During the second illness, at the instance of his medical attendants, the drains were examined. It was found that a recently constructed w.c. had been connected with an old square drain passing under the dining-room, and that this drain was blocked in consequence of a grating placed at its exit from the house in order to keep out the rats.

The second case was recently discovered in the house of a relative of Mr. H. B. Hewetson, of Leeds.

Illness led to investigation of drains and the discovery that they were blocked by a grating, the bars of which would barely admit a knife between them. This had been placed in the drain as a protection against "vermin."

XVI.

Obstructed drains in new houses.

(Communicated by Dr. Veale, Harrogate).

'I entered my present house in April, 1878. The house was a new one, semi-detached, the adjoining one being unoccupied. In July my wife complained to me that about 4 o'clock every afternoon she smelt sewage. At first I took no notice of her complaint, feeling assured that nothing could so soon be at fault, but shortly afterwards some of my children having occasional attacks of diarrhœa, I requested my landlord, a very intelligent man, a plumber, to examine my drain. On doing so, we found the main drain completely blocked about 5 or 6 yards from the house. On asking my landlord how he could account for the blocking so soon after occupation he said :—" The only cause I can assign for it is that during the time workmen are in a house they use wood shavings at the closet instead of paper, and as the house on completion was not occupied at once, the shavings, for want of water, lodged, became dry and hardened, and being insoluble, formed an obstruction." '

XVII.

Bell-wire tubes as conductors of sewer gas.

Professor Corfield tells me that on more than one occasion he has met with this condition.

XVIII.

A County Infirmary.

I cannot refrain from giving the following graphic account, by Mr. W. D. James, of Sheffield, of the conditions discovered, not many years ago, in one of our County Infirmaries; conditions in many respects no doubt, until recently, to be found in not a small number of our old hospitals; and to be found, I strongly suspect, even at the present time, in one of the County Asylums. In this Asylum outbreaks of erysipelas were still rife a short time ago, but out of tender regard for the feelings of the Surveyor, who is responsible for the state of the drains, the authorities refused to institute an investigation by a competent and independent inspector.

The letters A B C refer to a ground plan which accompanied the letter.

"I was house surgeon at the — Infirmary a few years ago, and for the greater part of the time suffered from sudden attacks of sore throat, accompanied by great prostration. These always came on in the night when I had gone to bed quite well. The nurses were always ailing in the same way. In eighteen months we had three distinct epidemics of erysipelas, once having to close the accident wards and suspend all operations. On examination every w.c. was defective, some joints being made of red lead spread on brown paper and put round the leaking pipe. After repeated appeal on my part, and two or three tinkering attempts to remedy the evil, the sewage matter came up as backwash as far as the grate at A. Then the authorities consented to spend money. The drain was opened at B (in the plan). I found it running under the building, an ordinary square stone drain, laid without any fall whatever. It was choked with sewage. I found there existed a machine for clearing it, an iron chain of which each link was a yard long, armed at one end with a kind of scraper.

This was thrust up the drain link by link, and worked about until the drain was thought to be clear. I saw an exactly similar machine used for clearing some of the town drains, which were in the same state as the one I speak of. The sewer was then traced under the wood cellar beneath the Infectious Diseases Wards to C, whether manhole or catchpit I could not determine, as the drain ran in at one side and out at the other close to the bottom. As far as C' the drain pipes, which were proper glazed pipes from exit from building, were choked with a solid mass of sand, hair, a tea cup, (fact!) mixed with ordinary sewage, and so firmly were they filled that on being taken up, a mould of the pipes could be shaken out. From C' to C" no drain whatever had been laid. From C" a good drain ran with a fall of many inches per yard into the main sewer. There were no ventilating pipes, no w.c. was efficiently trapped, no pipes from housemaids' sinks were trapped at all. The grate A was close beneath the outer door of the Accident Ward. The "new drain" from C to main sewer had been down ten years.

In addition to all this, the Deadhouse was underneath the Infectious Diseases Wards, within ten yards of the windows of which were the piggeries. In the summer months every drop of water used in the house had to be carried by hand from a well sunk in the Killas (an igneous rock) in the very middle of the building.

Many of these defects are now altered. As a commentary, I had these repeated throat attacks, and my successor died in six weeks, after two days' illness, from laryngitis!

XIX.
Gibraltar.—A conversation,

(Communicated by a lady).

" Mrs. B, my Gibraltar friend, called yesterday, and told me all about her husband's illness, which was typhoid, as I expected. He was nearly at death's door. I give you part of our conversation.

Mrs. A : Did you find anything wrong with the house?

Mrs. B (not understanding) : Wrong with the house! Why, it's one of the nicest houses in Gibraltar.

Mrs. A : Oh yes, but did you have the drainage inspected with a view to discovering the cause of the disease?

Mrs. B : Oh no; you see we came away in such a hurry.

Mrs. A : But you surely do not intend going back to it without having an inspection made and getting things put right?

Mrs. B : Put right! Why, my dear Mrs. A, there cannot be much wrong, for the drain is always bursting, then everything has to be carried away.

Mrs. A : Bursting! What makes it burst?

Mrs. B : Oh, because it can't get away! All the drainage is bad at Gibraltar.

Mrs. A : But perhaps you have no drains at all; and it is a cesspool that bursts.

Mrs. B : Oh no; we have a drain from the house which goes to the main sewer in the middle of the road, but it takes a sudden bend, where everything stops, then it all bursts up near the kitchen door. There really cannot be much wrong (she repeated this for the second time), for it bursts about once a month.

This is a good illustration of unconscious ignorance.

The dear good lady wept over her husband's sufferings, and dwelt on the narrow escape she had had of widowhood, and laughed over the bursting of the drain, and would not allow her mind to dwell on the possibility of anything being wrong there."

XX.

Testing for leakages of sewer gas in pipes and drains.

I had hoped to be able to lay down some definite rules on this subject, but, as will have been seen, the kinds of defect and points of leakage are so numerous that it is not easy to define a test that shall be always easily applicable.

Some, however, of the most readily available methods may be mentioned, such as—

A.—Those which appeal to the eye,

ex. gr., the flame of a lighted taper held over an untrapped pipe, or defective joint, or at the crevices in the floor covering a defective drain. (*Vide* plate III.)

Mr. Wheelhouse, of Leeds, tells me of the successful use of burning straw. A drain was opened outside a house and straw was burnt inside the drain, and in a short time the smoke was discovered at many points in the house, drawn by the fires through defects in pipes and drains. On the subject of "smoke testing," as practically carried out by sanitary engineers, Mr. W. P. Buchan, of Glasgow, has most kindly given me the following information :—" For the last five years I have used 'Watts' Vermin Asphyxiator."[*] The indiarubber is inserted into the top of a Buchan's trap, whence the smoke proceeds along the inside of the drain, and up the soil-pipe, and out above the roof. If leakages exist in the drains, &c., the smoke may come out at these in the house, in some cases being seen quite plainly issuing from the holes. The materials I have hitherto been using for testing are waste which has been used to wipe machinery, and sulphur, so that we both see the smoke and feel the smell of sulphur in the house tested if drains, &c., are leaking. In some cases one may be walking through clouds of smoke after a minute or so of application of the machine."

---

[*] John Watts & Co., Broad Weir Works, Bristol.

B.—Those which appeal to the sense of smell,
ex. gr., by pouring in at one or more points of the supposed defective drain some substance of powerful odour easily diffusible, the leaky points may be detected by the strong smell. Such are, Ether, the vapour of which is however highly inflammable—Oil of mint of the cheap variety, convenient, because it is very effective even in small bulk— Crude petroleum and common gas liquor useful for testing communications of main drains with buildings. "The plan usually adopted is to pour down each manhole about 15 gallons of the gas liquor at the highest point of each main drain. Then on examination of the drains from sinks, baths, and w.c's., the smell readily betrays the communication. The crude tar oil is explosive if a light be applied. Gas liquor is not inflammable."

(DR. CRICHTON BROWNE.)

APPENDIX.

Extracts from bye-laws with respect to new streets and buildings, issued by the Council of the Borough of Leeds, and allowed by the Local Government Board, July 12, 1878 :—

§ 4. No person shall construct any foundation of a new building on a site which has been previously used as a place for depositing night soil, refuse, or any offensive material which may have rendered such site liable to cause buildings erected thereon to be unhealthy, until such refuse or offensive material shall have been removed to the satisfaction of the Corporation, and such site shall not be built upon until the same shall have been approved by the Corporation.

§ 33. The person erecting any new building shall, as regards the construction of the drains of such building, comply with the requirements hereinafter specified, namely :—

(a.) He shall cause such building to be provided with sufficient drains to carry away the whole of the waste water and drainage from such building, and with suitable and sufficient spouts and fall pipes for conveying the rain water from the roof of such building to the drains.

(b.) He shall construct the lowest storey of such building at such a level as will allow of the construction of a sufficient drain from such building with an adequate fall in such drain, and so that such drain shall communicate with any sewer into which it may discharge, at a point in the upper half section of such sewer.

(c.) If there be no sewer within a distance of 100 feet from such building, he shall cause the drains to be taken to a cesspool properly constructed in accordance with these Bye-Laws.

(d.) He shall cause the drains of such buildings to be constructed of good glazed stoneware pipes or pipes of other equally suitable material; to be not less than 6 inches diameter for waste water and water-closet drains, and of not less than four inches diameter for rain water drains, to be laid with a proper fall and with water-tight socketted or other suitable joints.

(*e.*) He shall cause the lowest cellar, or basement storey, to be provided with a suitable and sufficient drain for the effectual drainage thereof.

(*f.*) He shall not construct any drain so as to pass under any building, except in any case where any other mode of construction may be impracticable, and in that case he shall cause such drain to be laid in the ground at such a depth that there shall be in every part a distance equal at the least to the full diameter of the drain, between the top of such drain and the finished surface of the ground, and he shall cause such drain to be laid in a direct line for the whole distance beneath such building, and to be embedded in and surrounded with good and solid concrete at least six inches thick all round.

(*g.*) He shall in the case of any back-to-back house, which is unprovided with any open space appurtenant thereto, cause the inlet to the drain or drains from such house to be at a point, as near as may be practicable to any external wall of such house, and he shall cause such inlet to be provided with a suitable trap.

He shall cause every pipe for conveying waste water from such house to the drain, to discharge immediately into the trap.

He shall also cause such waste pipe to be of lead or iron, and of not less than two inches diameter interior measurement.

(*h.*) He shall not construct in the drains any right angled junction, whether vertical or horizontal, but he shall cause every branch or tributary drain to join another drain obliquely in the direction of the flow of such drain.

(*i.*) He shall not allow any inlet to any drain to be made within any building, except such inlet as may be necessary from the apparatus of any water-closet, and he shall cause the waste pipe from every sink, bath, or lavatory, the overflow pipe from any cistern and every pipe for carrying off waste water, to be furnished with a syphon trap, and to be taken through an external wall of such building, and to discharge in the open air over a channel leading to a trapped gulley grating. Provided that the requirements of this clause shall not apply in the case of any back-to-back house, which is unprovided with any open space appurtenant thereto.

(j.) He shall in every case cause the drains to be furnished with a shaft from the exterior drain, not less than two inches and a half in diameter, communicating with the outer air above the eaves spouts.

(k.) He shall cause the drains to be efficiently trapped at some point near to their outfall, and he shall cause suitable and sufficient means of ventilation to be provided in such drains. He shall also cause every inlet to such drains, except such as may be provided for the ventilation thereof, to be properly trapped.

§ 34. Before commencing the erection of a new building in any street, the Owner or Builder shall, if there be a main sewer or drain within 100 feet of the site of such new building, make a connecting drain or sewer from such site to such main sewer or drain at such a depth as to carry off from the lowest excavations for a basement of such new building all the water capable of being carried off by such sewer or drain, and shall thereby or otherwise prevent such water from flowing into the basement of cellars of any adjoining or neighbouring buildings or into the walls thereof.

§ 35. No person shall construct a Cesspool in any case where an accessible outlet sewer is situated within 100 feet from the dwelling-house or building to be drained.

§ 36. Every person who shall construct a Cesspool in connection with a building shall construct such Cesspool at a distance of 15 feet at the least from a dwelling-house or public building, or any building in which any person may be, or may be intended to be employed in any manufacture, trade, or business,

§ 37. A person who shall construct a Cesspool in connection with a building shall not construct such Cesspool within the distance of 18 feet from any water supplied for use, or used, or likely to be used by man for drinking or domestic purposes, or for manufacturing drinks for the use of man, or otherwise in such a position as to endanger the pollution of any such water. Provided always that the foregoing requirements shall not apply where such water is supplied by the Corporation and conveyed in metal pipes.

§ 38. Every person who shall construct a Cesspool in connection with a building, shall construct such Cesspool in such a manner and in such a position as to afford ready means of access to such Cesspool, for the purpose of cleansing such Cesspool and of removing the contents thereof, and in such a manner and in such a position as to admit of the contents of such Cesspool being removed therefrom, and from the premises to which such Cesspool may belong without being carried through any dwelling-house or public building, or any building in which any person my be or may be intended to be employed in any manufacture, trade, or business.

He shall not in any case construct such Cesspool so that it shall have, by drain or otherwise, any outlet into or means of communication with any sewer.

§ 39. Every person who shall construct a Cesspool in connection with a building, shall construct such Cesspool of good brickwork in cement properly rendered inside with cement, and with a backing of at least 9 inches of well puddled clay around and beneath such brickwork.

He shall also cause such Cesspool to be arched or otherwise properly covered over, and to be provided with adequate means of ventilation.

§ 40. Every person who shall construct a Water-closet or Earth-closet in a building shall construct such Water-closet or Earth-closet in such a position that one of its sides, at the least, shall be an external wall.

§ 53. Every person who shall construct a water-closet in connection with a building used or intended to be used as a dwelling-house or shop, shall cause such water-closet to be provided with a 4-inch internal diameter soil pipe of lead or iron, which shall be continued upwards without diminution of its diameter and (except where unavoidable) without any bend or angle being formed in such soil pipe to such a height, and in such a position as to afford, by means of the open end of such soil pipe, a safe outlet for sewer air.

§ 58. Every person who shall intend to let for occupation, or being the owner thereof, shall occupy as a dwelling-house any new building of which the rateable value is under £20, shall give seven clear days' notice thereof to the Corporation. Such notice shall not be given until the building is actually completed, and shall be delivered at the office of the Building Inspector of the Sanitary Authority, at Leeds, and such building shall not be occupied as a dwelling-house until the drainage thereof has been made and completed, or until such building has after examination been certified by the Surveyor to be fit for human habitation, and the Surveyor shall give a certificate to that effect if he is satisfied after examination that such building is fit for human habitation.

§ 65. Such person shall also, before proceeding to cover up any sewer or drain, or any foundation of a Building, deliver or send, or cause to be delivered or sent to the said Surveyors, two days' notice in writing, in which shall be specified the date on which such person will proceed to cover up such sewer, drain, or foundation.

INDEX.

The Roman numerals refer to the Plates, the small figures to the pages of descriptive letter-press.

Additions to house, built over forgotten drains, XXXVII., 75
Age of leaden soil-pipes, danger from, XXIV., 49
Air, fresh, how to admit into carriages, LXX., 143; into rooms, LXVIII., 137, III., 5.
—— from drains circulating through house, III., 5 (see also *Sewer-gas*)
——, provisions for admission of, III., 5, LXVIII., 137
——, quantity required to feed chimney, III , 5
Air-brick giving passage to sewer-gas, 148
Air-grate for passage of air through soil-pipe, XX. (2), 41
Angles, pipes joined at, cause of leakage, XLVI., 93
Arsenical wall-papers, their dangers, LXVII., 135
Ash-pit, refuse from, for mortar and plaster, LV., 111

Basin and syphon to replace pan-closet, XXI. (2), 43
Bath, pipes from, badly arranged, VII. (A, B, C), 13, I., 1
——————, properly arranged, VII. (D), 13, II., 3
——, waste-pipe into drain cut off and left open, XV., 29
——————, opening untrapped into fall-pipe, XVI., 31; into soil-pipe, I., 1
Bedroom-window, sewer-gas passing into, from fall-pipe, XVII., 33
Bedrooms, passage of sewer-gas into, XIII. 25, IX. 17, XI., 21, XII., 23, XV , 29, XVI., 31, XVII. 33, XVIII., 36, XX., 39, XXIV., 49, XXX., 61, LXII., 125, XXXI., 63, XL 81, XLIX., 99, LXIII., 127, 147, 148
Beer tainted by sewer-gas in larder, XXVI., 53
Bell-trap, illusory protection afforded by, L., 101
Bell-wire tubes as conductors of sewer gas, XVII., 153.
Bend properly formed in drain-pipes, XLVI., 93
——, sudden, causing bursting of pipes, 156
"Black damp" caused by escape of sewage, XXIV., 49
Boiler, water in, fouled by sewer-gas, XIV., 27
Brougham, window-ventilator in roof of a, LXX., 143
Burning straw, smoke from, to detect leakage of sewer-gas, 157
Bursting of drains at Gibraltar, 156
Butler falling into unsuspected cesspool in wine-cellar, XXXVIII., 77
Bye-laws of the Borough of Leeds, extracts from, 159

Candle-flame at key-hole, lessons to be learnt from, III., 5
—————— for detection of flaws in joints, XLV., 91, 157
Carr, Mr H., on arsenical wall-papers, 135

Cellars damp from slop water, LXI., 123
———, sewage in, XXXVI., 73, 150; LIV. 109
Cesspool below bedroom-window, 147
——— -, chinks in side of, LXIV., 129; faulty to laundry, LXIV., 129; formed by leaking pipes, XLVI., 93; by obstructed pipe, LXIII., 127; by termination of drain-pipe, 73; formed in disused well under house, XXXIX., 79; formed under hall by leaking stone drain, XL., 81
——— - overflowing into basement of house, XXXVI., 73, 119; into rain-water tank, XXXI., 63; XXXV., 71; into well, XXXIII., 67
——— - under new dining-room, XXXVII., 75; unsuspected in wine-cellar, XXXVIII., 77; unventilated save into house or through fall-pipe, 42, XXXI., 63
Cesspools in Edinburgh, 119
———, Leeds Bye-laws regarding, 161, 163
——— under London houses, LIX., 119
Chimney, ventilating pipe of drain ending in, XVIII. (A), 35
——————————————————— near top of, XVIII. (B), 35
Churchyard, infiltration from, into cellars, LX., 121
Cistern for house distinct from w.c. cistern, II., 3
——— overflow pipe from, passing into open air, XX. (2), 41; into soil-pipe, XX. (1), 41, 43
——— supplying boiler, overflow pipe from, passing direct into drain, XIV., 27
Common sanitary faults of ordinary houses, I., 1
Connection, defective, of waste-pipe, XLIV., 89
Continental health resorts, their dangers, LXII., 125
Continental hotel, drain in central court of, 148
Cotton-wool for exclusion of dust in ventilation, LXIX., 141
County Infirmary, condition of drains in a, 154
Covering of drains, Leeds Bye-laws regarding, 163
Crowbar, tapping with, to discover spaces under stone floor, XXX., 61
Curves in drains made by straight pipes, XLVI., 93

Dairies, fissures leading to drains in floors of, for sweepings, XXVII., 56
——— Sanitary inspection of, why necessary, XXXIV., 69
Dangers from neighbours' drains, XLII., 85
Dead-house in Hospital, under infectious diseases wards, 155
Defective junction of pipes and drains, I., 1. (see also *Junctions*)
"Diffuser," (Harding's), for preventing draughts, LXVIII., 139
Dining-room over cesspool, XXXVII., 75
Diphtheria from sewage-emanations, XXXVI., 73, in houses with manure heap against wall, LXV., 131.
Dirt, how to exclude in ventilating rooms, LXVIII., 137
Disconnection of waste-pipes from drain-pipes, V., 9
Dish-stone in larder leading to drain, XXVI., 53
——— ——— scullery leading to rain-water tank, with overflow direct into drain, XXVIII., 57
D.ain carried across upper part of well, XXXIV., 69
——— ——— over instead of through a rock, LII., 105
——— ——— common stone, under tiled hall leaking and forming a cesspool, XL., 81
——— ——— communicating with rain-water tank, XXVIII., 57

Drain, damage to by rats runs, XXXII., 65
———, fall of, defective, XLIX., 99, LIII., 107
———, how to tap, XLVII., 95
——-- led up hill, XLIX., 99, LXIV., 129, 95
—— made of imperfect tubes, XLIV., 89, XLIII., 87
—— not connected with main sewer, 89, 149, 150
—— outside house, XX., 41, II., 3
·———, settling of, and leakage from, LXI., 123
—— terminating in soil and forming cesspool, 73
——- under house, and obstructed, LXIII., 127, 153, and with joints leaking, I. 1
—— ventilated into roof, I. 1, 146
Drainage at Gibraltar, 156
———- defects of, how influencing vaccination, LXVI., 133
Drain-pipes, defective, trade in, XLIII., 87; disconnected and misconnected, L. 101; imperfect and with unluted joints, XLIV., 89, XLIX., 99; leaking into well, XXXII., 65, XXXIX., 79; materials for joining, 117.
Drain-work, how "scamped," XLIII., 87, XLIX., 99, XLIV., 89.
Drains, obstructions in (see *Obstructions*) how to test for, LXIV., 129
—— of buildings, Leeds Bye-laws regarding, 159
——— of house sealed-up during making of new sewer, 149
———, search for, in absence of plans, LVII., 115
Draughts, irregular, in rooms, how to stop, III., 5, LXVIII., 137
Drawn lead-pipes, their advantages, XXIV., 49
Dust, how to exclude from glass-cases, LXIX., 141

Earth-closets in buildings, Leeds Bye-law regarding, 162
Edinburgh, old cesspools under houses in, 119
Erysipelas due to bad drainage, XVI., 31, XLIX., 99, 152, 153
————- attending vaccination, due to cesspool under window, LXVI., 133
Ether poured into drain to detect leakage in drain-pipes, 158
Evaporation of water in traps, a cause of escape of sewer-gas, IX., 17
Exclusion of dirt from air supplied to rooms, LXVIII., 137
Eyelet for new junction in pipe already laid, XLVII., 96

Fall, defective, of drain, XLIX , 99, LIII., 107, I., 1, 153
Fall of pipe from water-tank to sewer in wrong direction, XXXV, 71
Fall-pipe communicating directly with the drain, XVI , 31, XVII., 33
———— conducting foul air from tank, I., 1, XXXI., 63
———— discharging into gulley, II., 3
False roof, w.c. ventilated into, 146; soil-pipe ventilated into, 146
Fires, effect of, in drawing air from drains, IV., 7
Flame of candle, as indicating source of air supplied, III., 5
——————— to detect leakage of sewer-gas, 157
"Floral Art Ventilator," 140
Foundations, sinking of, a cause of open pipe-joints, XLI., 83
Freshness of atmosphere in rooms, means for securing, LXVIII., 139

Gas liquor for testing for leakages in drain-pipes, 158

Gibraltar, condition of drains in, 156
Glass-cases, exclusion of dust from, LXIX., 141
Grates, open, in cellar, for purpose of swilling floor, XXVI., 53
Gratings in drains obstructing flow, 152
Graveyard, infiltration from, into cellars, LX., 121
Gullies, description of, V., 9
—————, Leeds Bye-laws relating to, 161
—————, their advantages, II., 3, V., 9, 31

Hall, cesspool formed under, by leaking stone drain, XL., 81
"Harding's Diffuser," for preventing draughts, LXVIII., 139
Highland Shooting Lodge, defects of drainage in, LXIV., 129
Holes in pipes for junctions, XLVII., 95
————————— caused by rats, XIX., 37, 151
Hot-bed against side of house, ill effects of, LXV., 131
Hotel, Continental, drains in, 148
House costing £3000, but with drains of "seconds" pipes, 89
————— drain, proper position of, II., 3
————————— under floor of room, I., 1; under house, XLIX., 99; LI., 103
————— with every sanitary arrangement faulty, I., 1
————————— faulty sanitary arrangements avoided, II., 3
Houses built on unhealthy rubbish heaps, LVI., 113
—————, necessity for plans of drains of, LVII., 115
Housemaid's sink, untrapped, and discharging into soil-pipe, XIII., 25, I., 1
———————————, water from, keeping cellar damp, LXI., 123

Illness after vaccination, from drainage faults, LXVI., 133
————— caused by old papers remaining on walls, 135; from arsenical wall papers, LXVII., 135
————— instances of, due to sewage emanations, LII., 105, LX., 121, LXI., 123, 99, 103, 105, 146, 147, 149, 151, 153 (see also *Typhoid Fever*, *Erysipelas*, etc.)

"Jerry builders" and their dealings, XLIII., 87; LV., 111; LVI., 113
"Jerry veal" and its analogies in the building trade, XLIII., 87
Joints, imperfect, of soil-pipe allowing escape of sewer-gas into house XX. (1) 39, XLV., 91; of fall-pipe leading to drain, XVI., 31, 148
————— of drain gaping, from sinking of foundations, XLI., 83, XLII., 85, LXI., 123
————— of soil-pipe made of brown paper and red lead, 153
"Junctions" badly made, XLVII. (A), 95, L., 101; proper, XLVII. (B), 95; broken pipes at, XLIX., 99, XXXIX., 79; complete absence of, LI., 103; defective, of ventilating pipe with soil-pipe, 147

Keeping cellar, untrapped sink-stone in, XXV., 51
Kitchen and larder fouled by sewage-emanations, LXII., 125, XXXVI., 73
—————, passage of sewer-gas into, through overflow-pipe of cistern. XIV., 27
Kitchen-sink trapped and disconnected from drain, II., 3
————————— untrapped, VI., 11, I., 1, 31

Lady's, A, method of testing and inspecting drains, LXIV., 129
Lavatories and baths, defects in, and how to remedy, VII., 13
Lavatory in bedroom trapped, but discharging into soil-pipe of W.C., XII., 23
——, overflow of, joining waste-pipe below trap, X., 19
——, waste-pipe of, passing untrapped into drain-pipe, XI., 21, I., 1
——— ————————, properly trapped, II., 3
Lead-pipes, age of, XX., 39 ; eaten through by rats, XIX., 37 : by sewer-gas, XX., 39 ; joined with putty, XLV., 91 ; seams in, XX., 39
Leakage into soil from pipes laid the wrong way, XLVIII., 97
—— of sewer-gas in pipes and drains, how to test for, 157, XIX., 37, III., 5, XLV., 91, XXIV., 49
Leeds, Borough of, extracts from Bye-laws of, 159
Lime from tan-pits used for making mortar, LV., 111
London houses, old cesspools under, LIX., 119
Luting, drain-pipes without, XLIX., 99 ; lead-pipe passed into drain-pipe without, XLIV., 89

Manure, soakings from, entering wells, XXXIV., 69
Manure-heap against house wall, LXV., 131
Mayfair, discovery of pit containing remains of cattle under house in, 119
Meat tainted by sewer-gas in larder, XXVI., 53
Milk made poisonous by use of contaminated well-water, XXXIV., 69
Milk tainted by drains of dairy, XXVII., 55
——————— sewer-gas in larder, XXVI., 53
Mortar made from lime used in tan-pits, 111 ; from road-scrapings and midden refuse, LV., 111
Museums, how to exclude dust from cases in, LXIX., 141

Neighbours' drains, dangers from, XLII., 85
New drains not connected with sewer, 150, LI., 103
New made ground, drains in, precautions regarding, XLI., 83
Nursery, sewer-gas in, LXVI., 133, 146

Obliteration of old drains and cesspools, necessity for, XXXVII., 75
Obstructed drains in new houses, 153.
Obstruction to flow of sewage, from carrying 12-inch pipes into 2-inch ones, 109 ; from improper junctions, XLVII., 96, XLIV., 89, L., 101 ; from interpolating 6-inch pipe between two smaller ones, LIV., 109 ; from old wall-papers being thrown into drain, 149 ; from roots in pipes, LVIII., 117, LXIV., 129 ; from sand, hair, &c., 154 ; from settling of pipes, LXI., 123 ; from sudden bends, 156 ; from taking curved tubes over rock, LII. (A), 105 ; from vermin traps or grates, 152
Occupation of new houses, Leeds Bye-laws regarding, 163
Oil of mint poured into drains to detect leakage in pipes, 158
Open drain-joints, from giving way of foundation, XLI., 83, XLII., 85 ; from imperfections of tubes, XLIV., 89; from pipes laid the wrong way, XLVIII., 97 ; from putty-joints, XLV., 91 ; from using straight tubes for curves, XLVI., 93; from want of luting, XLIX., 99 ; how to detect, XIX., 37, 157
Outlet for sewer-air from W.C., Leeds Bye-law regarding, 163
Overflow-pipe from cesspool at higher level than inlet, LXIV., (2) 129
——— ——— of bath into soil-pipe VII. (A, B), 13, I., 1
——— ——— ——————— open air, VII. (C, D), 13, II., 3

Overflow-pipe of cistern into drains, XIV., 27, XX., 41, IX., 17. I., 1, 33
———— ———————— open air, II., 3

Pan-closet, XXI., 43, XX., 39, I., 1
———— —— substitute for, XXI., 43, XX., 39
Paper containing arsenic on walls of room, LXVII., 135
——,——, old, on walls of room, 135
Petroleum for testing for leakages in drain-pipes, 158
Piggeries close to Hospital windows, 155
Pipe, 6 inch between two smaller pipes, LIV., 109
Pipes, damaged, used for drains, XLIV., 89
———— laid with flange downhill, XLVIII., 97
Plan of drains seldom possessed by owner or tenant, LVII., 115
Plaster made from lime from tan-pits, 111 ; from road-scrapings and midden refuse, LV., 111
Puerperal fever due to defective drainage, LXIII., 127
Putty-joints in leaden soil-pipes, XLV., 91, XX., 39

Rain-water tank under floor, I., 1
———————————— with overflow direct into drain or cesspool, XXVIII., 57, XXXI., 63, XXXV., 71, I., 1
———— ———— ———— without overflow-pipe, XXIX., 59
Rats, appearance of, in kitchens, lesson to be learnt from, XIX., 37, 89
———, damage to leaden-pipes, caused by, XIX., 37,
———, runs made by, causing drain-joints to open, XXXII., 65
Road-scrapings for mortar and plaster, LV., 111
Roof-light and ventilation for a brougham, LXX., 143
Roots of trees blocking drains, LVIII., 117, LXIV., 129
Rubbish-heaps, houses built on, LVI., 113

Sanitary inspection of dairies, necessity for, XXXIV., 69
Save-all tray beneath w.c., with untrapped waste-pipe, XXII., 45
———— ———— of bath-taps untrapped and passing into soil-pipe, VII. (C), 13, I., 1
———— ———— ———— ———— properly trapped and passing into gulley, II., 3 VII. (D), 13
Scamped drain-work, XLIV., 89, XLV., 91, XLVI., 93, LI., 103, L., 101, XLIX., 99, XLVII., 95, LII., 105, LIII., 107, LIV., 109
Screen in ventilator for cleansing air, LXVIII., 137
"Scribner's Monthly Magazine" remarks in, on pan-closet, 43 ; on sanitary condition of New York, 113
"Seconds," or defective drain-pipes, use of, XLIII., 87, XLIV., 89
Sewage, escape of, from defective fall, XLIX., 99, LIII., 107
————, how it may gain access to water, XXXII., 65, XXXIII., 67, XXXIV., 69
Sewage, liquid, escape of, through defective junction, XLIX., 99 ; through old leaden soil-pipe, XXIV., 49
————, passage of, into disused well, XXXIX., 79 ; into rain-water tank, XXXI., 63, XXXV., 71
————, saturating soil under cellar, LI., 103; under kitchen, XLIV., 89, 93 ; under ground-floor, LII., 105, LXII., 125, I., 1, XXIV., 49, XXXVI., 73, XL., 81, XLI., 83, XLIX., 99

Sewer-gas, diffusion of, through walls, XLII., 85
——————, escape of, from pipes, how to detect, 157, XIX., 37, XXIV., 49, III., 5
——————, sometimes due to evaporation of water in syphon-trap, IX., 17
——————, through defective joints in fall-pipe, XVI., 31; in soil-pipe, XX. (1), 39, XLII., 85, XLIV., 89, XLV. 91
—————— passing horizontally to a room 22 feet distant, 151
—————— into boiler, through over-flow pipe of cistern, XIV., 27; into chimney through ventilating pipe of drain, XVIII., 35; into false roof, 146
—————— into house, from cesspool formed under hall, XL., 82; under dining-room, XXXVII., 75; near bedroom window, 147; from disused well near defective drains, XXXIX., 79; from rain-water tank connected with drain, XXX., 61
—————— into house, through air-brick near defective joint of fall-pipe, 148; through chimney from ventilator of drain, XVIII., 35: through unconnected soil-pipe, LI., 103; through dish-stone in larder, XXVI., 53
—————— into house through fall-pipe and ventilator of soil-pipe opening near bedroom window, XVII., 33; through holes in soil-pipe, XXIV., 49; through kitchen sink-pipe, IV., 7; through overflow-pipe of cistern, XX., 41; of lavatory, X., 19; of rain-water tank, XXXI., 63; through untrapped sink-pipe, XIII., 25 I., 1; through untrapped waste-pipe of lavatory, XI., 21, I., 1, LXVI., 133
Sink, discharging into fall-pipe, feeding tank under room, 59; into soft-water cistern under cellar floor, XXIX., 59
—————— over grating of untrapped drain, XXV., 52
——, waste-pipe, from, cut off and left open, XV., 29
Sink-pipe passing into open mouth of drain-pipe, XIX., 37, XXV, 51
Sinkstones in cellars, their dangers, XXVI., 53
Sites for buildings, Leeds Bye-laws regarding, 159
Sitting-room, soil-pipe in corner of, XXIII. 47
Soil-pipe broken at junction with drain, XX., 39, XXXIX., 79
—————— communicating with upper rooms of house, 146
————, holes in, from age, XXIV., 49
—————— inside house, XX., 39, I., 1, XXIII., 47, XXIV., 40
—————, joints of, defective, XX. (1), 39, I., 1, XXXII., 65, XXXIV., 69, XL., 81, XLII., 85, XLIV., 89, XLV., 91, XLVI., 93
—————— missing drain-pipe and discharging contents below ground-floor, LII., 105
—— —— not ending in sewer, but terminating in mass of solid rock, LI., 103
—— —— outside house, II., 3, XX. (2), 41, XXI., 43
Soil Pipe terminating above eaves XX., (2) 41.
————— unventilated, XX. (1) 41
————— ventilated into false roof, 146; into house, 147
————— ventilator of, terminating just below bedroom window, XVII. (B) 33
Sore throat from sewage-emanations, XXXVI., 73, IV., 7
Sounding floor of cellars, necessity for, XXX., 61, XXXVIII., 77
Speculating builders: their bargains, XLIII., 87, XLIV., 89, LV., 111
Syphon sanitary basins for w.c., XXI., 45, XX., 30
Syphon-trap, effect of absence of, IV., 7
—————— —————— to cut off sewer-gas from soil pipe, XX. (2), 41

Tank connected with w.c. overflowing into room, LXII., 125
Tea cup, a, obstructing drains, 155
Testing for leakages of sewer-gas, methods of, 157, XIX., 37, XXIV., 49, XLV., 91
Tobin's ventilating tube, LXVIII., 137
Traps, disused, dangers from, IX., 17
────── unsyphoned, how produced and remedy for, VIII., 15
Typhoid fever at Gibraltar, 156
────────── from drinking water polluted by typhoid discharges, XXXII., 65; polluted by sewage, XXXIII., 67, XXXIV., 69
──────────── emanations from tank connected with w.c., LXII., 125
────────────────────────── receiving washings of sink, XXIX., 59
──────────── leaking drains, XLII., 85
──────────── milk kept in foul dairy, XXVI., 53, XXVII., 55
──────────── sewer-gas from cesspool overflowing into basement, XXXVI., 73
────────────────── through water-traps, 42
────────────────── under dining room, XXXVII., 75
──────────── from obstructed pipes, LIV., 109
──────────── soil pipe, 147
──────────── unsuspected tank connected with drain, XXX., 61.
──────────── through fall-pipe, XVII., 33
──────────── untrapped waste-pipe, L., 101
──────────── ventilating pipe of drain ending in chimney, XVIII., 35

Vaccination and drainage faults, LXVI., 133
Ventilating shaft ending in chimney, XVIII. (A), 35
────────── near top of chimney, XVIII. (B), 35
────────── window, XVII. (B), 33
Ventilating tube on drain side of syphon-trap, XX. (2), 42
Ventilation, how to effect at small cost, III., 5; of carriages, LXX., 143; without dirt, LXVIII., 137
────── of soil-pipes, II. 3, XVII., 33, XX. (2), 42, I., 1, LXIV., 129, 163
Ventilator, Floral Art, 140
Ventilators, passive, conditions for efficiency of, 140
Vicarage, cellars in, "standing in water," LXI., 123
────── rendered unhealthy by adjoining grave-yard, LX., 121
Villages, drinking water in, XXXII., 65

Wall blackened by sewage, XXIV., 49
──── soaked through by manure, LXV., 131
Wall-papers, containing arsenic, illness from, LXVII., 135
────── thrown into and obstructing drains, 149
Walls of rooms, old paper remaining on, LXVII., 135
──────, passage of sewage through, XLII., 85, XLVI., 93
────── plastered with mud and midden refuse, LV., 111
Waste-pipe of bath and sink cut off, pipe open, XV., 29

Waste-pipe of bath trapped, II., 3, VII., 13; untrapped, I., 1, XVI., 31, LXVI., 133
―― ――― of kitchen sink, untrapped, IV., 7, L., 101
―― ― ― of lavatory untrapped, I., 1, VII. (A), 13, LXVI., 133
―― ― ― projecting into drain-pipe so as to cause obstruction, XLIV., 89
Waste-water from sink-pipe discharged into soft water cistern, XXIX., 59
――― ―― ―――― ――― ――― ―― untrapped drain, XXV., 51
Water from boiler contaminated by sewer-gas, XIV., 27
―― closet, faulty position of, XX. (1), 39, I., 1, XL., 81, LIX., 119; improved, XXI. (2), 43; proper position of, II., 3, XX. (2), 39; old, in centre of house, XXIV., 49; six in centre of house, LXIII., 128; ventilated into false roof, 146
―― ―― ―― cistern, with overflow into soil-pipe, I., 1
Water-closet, faulty, XX. (1), 39; XXI. (1), 43, I., 1
―― ―― with the faults remedied, XX. (2), 41, XXI. (2), 43, II., 3
Water-closets, state of, in a County Infirmary, 153
――――――, position and construction of, Leeds Bye-laws regarding, 162, 163
Water-tank, disused and unsuspected, under cellar floor, XXX., 61
―――― tightness of joints reduced by improper laying of pipes, XLVIII., 97
Well in or near farmyard receiving soakings from manure, XXXIV., 69
―― polluted by sewage, XXXII., 65, XXXIII., 68, XXXIV., 69
―― (disused) under house fouled by leakage from defective drain, XXXIX., 79
Wells, with drains carried through side of or across, XXXIV., 69
Willow roots obstructing drains, LVIII., 117, LXIV. (A), 129
Window-ventilator in roof of a brougham, LXX., 143
Windows, admission of foul air into, from fall-pipe, XVII. (A), 33, I., 1
―― ― ――― ―――― ――― ventilator of soil-pipe, XVII.(B), 33
Wine-cellar, unsuspected cesspool in, XXXVIII., 77
Wire with baize, for admission of air and exclusion of dust, LXIX., 141

CHARLES GOODALL, PRINTER, COOKRIDGE STREET, LEEDS.

September, 1896.

MEDICAL

AND

HYGIENIC WORKS

PUBLISHED BY

D. APPLETON & CO., 72 Fifth Avenue, New York.

AMERICAN GYNÆCOLOGICAL AND OBSTETRICAL JOURNAL. (Monthly.) Edited by J. Duncan Emmet, M.D., and J. N. West, M.D. $4.00 per annum ; single copy, 35 cents.

AULDE (JOHN). The Pocket Pharmacy, with Therapeutic Index. A *résumé* of the Clinical Applications of Remedies adapted to the Pocket-case, for the Treatment of Emergencies and Acute Diseases. 12mo. Cloth, $2.00.

BARKER (FORDYCE). On Sea-Sickness. A Popular Treatise for Travelers and the General Reader. Small 12mo. Cloth, 75 cents.

BARKER (FORDYCE). On Puerperal Disease. Clinical Lectures delivered at Bellevue Hospital. A Course of Lectures valuable alike to the Student and the Practitioner. Third edition. 8vo. Cloth, $5.00 ; sheep, $6.00.

BARTHOLOW (ROBERTS). A Treatise on Materia Medica and Therapeutics. **Eighth edition.** Revised, enlarged, and adapted to "The New Pharmacopœia." 8vo. Cloth, $5.00; sheep, $6.00.

BARTHOLOW (ROBERTS). A Treatise on the Practice of Medicine, for the Use of Students and Practitioners. **Seventh edition,** revised and enlarged. 8vo. Cloth, $5.00; sheep, $6.00.

BARTHOLOW (ROBERTS). On the Antagonism between Medicines and between Remedies and Diseases. Being the Cartwright Lectures for the Year 1880. 8vo. Cloth, $1.25.

BASTIAN (H. CHARLTON). Paralyses: Cerebral, Bulbar, and Spinal. A Manual of Diagnosis for Students and Practitioners. With 136 Illustrations. Small 8vo, 671 pages. Cloth, $4.50.

BASTIAN (H. CHARLTON). Paralysis from Brain Disease in its Common Forms. With Illustrations. 12mo, 340 pages. Cloth, $1.75.

BILLINGS (F. S.). The Relation of Animal Diseases to the Public Health, and their Prevention. 8vo. Cloth, $4.00.

BILLROTH (THEODOR). General Surgical Pathology and Therapeutics. A Text-Book for Students and Physicians. Translated from the tenth German edition, by special permission of the author, by Charles E. Hackley, M. D. **Fifth American edition, revised and enlarged.** 8vo. Cloth, $5.00; sheep, $6.00.

BOYCE (RUBERT). **A Text-Book of Morbid Histology.** For Students and Practitioners. With 130 Colored Illustrations. Cloth, $7.50.

BRAMWELL (BYROM). Diseases of the Heart and Thoracic Aorta. Illustrated with 226 Wood-Engravings and 68 Lithograph Plates—showing 91 Figures—in all 317 Illustrations. 8vo. Cloth, $8.00; sheep, $9.00.

BRYANT (JOSEPH D.). A Manual of Operative Surgery. **New edition, revised and enlarged.** 793 Illustrations. 8vo. Cloth, $5.00; sheep, $6.00.

BURT (STEPHEN S.). Exploration of the Chest in Health and Disease. 8vo. 210 pages. With Illustrations. Cloth, $1.50.

CAMPBELL (F. R.). The Language of Medicine. A Manual giving the Origin, Etymology, Pronunciation, and Meaning of the Technical Terms found in Medical Literature. 8vo. Cloth, $3.00.

CARMICHAEL (JAMES). Disease in Children. A Manual for Students and Practitioners. Illustrated with Thirty-one Charts. 12mo, 591 pages. (STUDENTS' SERIES.) Cloth, $3.00.

CASTRO (D'OLIVEIRA). Elements of Therapeutics and Practice according to the Dosimetric System. 8vo. Cloth, $4.00.

CHAUVEAU (A.). The Comparative Anatomy of the Domesticated Animals. Revised and enlarged, with the co-operation of S. Arloing, Director of the Lyons Veterinary School. Second English edition. Translated and edited by George Fleming, C. B., LL. D., F. R. C. V. S., late Principal Veterinary Surgeon of the British Army; Foreign Corresponding Member of the Société Royale de Médecine, and of the Société Royale de Médecine Publique, of Belgium, etc 8vo, 1084 pages, with 585 Illustrations. Cloth, $7.00.

CORNING (J. L.). Brain Exhaustion, with some Preliminary Considerations on Cerebral Dynamics. Crown 8vo. Cloth, $2.00.

CORNING (J. L.). Local Anæsthesia in General Medicine and Surgery. Being the Practical Application of the Author's Recent Discoveries. With Illustrations. Small 8vo. Cloth, $1.25.

DAVIDSON (ANDREW). Geographical Pathology: An Inquiry into the Geographical Distribution of Infective and Climatic Diseases. 2 vols. 8vo. Cloth, $7.00.

DENCH (E. B.). Diseases of the Ear. A Text-Book for Practitioners and Students of Medicine. With 8 Colored Plates and 152 Illustrations in the text. 8vo. Cloth, $5.00; sheep, $6.00.

DEXTER (FRANKLIN). The Anatomy of the Peritonæum. 12mo. With 89 colored Illustrations. Cloth, $1.50.

DOTY (ALVAH H.). A Manual of Instruction in the Principles of Prompt Aid to the Injured. Including a Chapter on Hygiene and the Drill Regulations for the Hospital Corps, U. S. A. Designed for Military and Civil Use. **Second edition, revised and enlarged.** 12mo. 121 Illustrations. Cloth, $1.50.

ELLIOT (GEORGE T.). Obstetric Clinic: A Practical Contribution to the Study of Obstetrics and the Diseases of Women and Children. 8vo. Cloth, $4.50.

EVANS (GEORGE A.). Hand-Book of Historical and Geographical Phthisiology. With Special Reference to the Distribution of Consumption in the United States. 8vo. Cloth, $2.00.

EWALD (C. A.). Lectures on the Diseases of the Stomach. By Dr. C. A. Ewald, Professor of Pathology and Therapeutics in the University of Berlin, etc. Translated from the German by special permission of the author, by Morris Manges, A. M., M. D. Cloth, $5.00; sheep, $6.00.

FLINT (AUSTIN). Medical Ethics and Etiquette. Commentaries on the National Code of Ethics. 12mo. Cloth, 60 cents.

FLINT (AUSTIN). Medicine of the Future. An Address prepared for the Annual Meeting of the British Medical Association in 1886. With Portrait of Dr. Flint. 12mo. Cloth, $1.00.

FLINT (AUSTIN, JR.). Text-Book of Human Physiology; designed for the Use of Practitioners and Students of Medicine. Illustrated with three hundred and sixteen Woodcuts and Two Plates. **Fourth edition, revised.** Imperial 8vo. Cloth, $6.00; sheep, $7.00.

FLINT (AUSTIN, JR.). The Physiological Effects of Severe and Protracted Muscular Exercise; with Special Reference to its Influence upon the Excretion of Nitrogen. 12mo. Cloth, $1.00

FLINT (AUSTIN, JR.). The Source of Muscular Power. Arguments and Conclusions drawn from Observation upon the Human Subject under Conditions of Rest and of Muscular Exercise. 12mo. Cloth, $1.00.

FLINT (AUSTIN, JR.). Physiology of Man. Designed to represent the Existing State of Physiological Science as applied to the Functions of the Human Body. Complete in 5 vols., 8vo. Per vol., cloth, $4.50; sheep, $5.50.
⁎ Vols. I and II can be had in cloth and sheep binding; Vol. III in sheep only. Vol. IV is at present out of print.

FLINT (AUSTIN, JR.). Manual of Chemical Examinations of the Urine in Disease; with Brief Directions for the Examination of the most Common Varieties of Urinary Calculi. Revised edition. 12mo. Cloth, $1.00.

FOSTER (FRANK P.). Illustrated Encyclopædic Medical Dictionary: Being a Dictionary of the Technical Terms used by Writers on Medicine and the Collateral Sciences in the Latin, English, French, and German Languages. The work consists of Four Volumes, and is sold in Parts; Three Parts to a Volume. (*Sold only by subscription.*)

FOURNIER (ALFRED). Syphilis and Marriage. Translated by P. Albert Morrow, M. D. 8vo. Cloth, $2.00; sheep, $3.00.

FREY (HEINRICH). The Histology and Histochemistry of Man. A Treatise on the Elements of Composition and Structure of the Human Body. Translated from the fourth German edition by Arthur E. J. Barker, M. D., and revised by the author. With 608 Engravings on Wood. 8vo. Cloth, $5.00; sheep, $6.00.

FRIEDLANDER (CARL). The Use of the Microscope in Clinical and Pathological Examinations. Second edition, enlarged and improved, with a Chromolithograph Plate. Translated, with the permission of the author, by Henry C. Coe, M. D. 8vo. Cloth, $1.00.

FUCHS (ERNEST). Text-Book of Ophthalmology. By Dr. Ernest Fuchs, Professor of Ophthalmology in the University of Vienna. With 178 Woodcuts. Authorized translation from the second enlarged and improved German edition, by A. Duane, M. D. Cloth, $5.00; sheep, $6.00.

GARMANY (JASPER J.). Operative Surgery on the Cadaver. With Two Colored Diagrams showing the Collateral Circulation after Ligatures of Arteries of Arm, Abdomen, and Lower Extremity. Small 8vo. Cloth, $2.00.

GERSTER (ARPAD G.). The Rules of Aseptic and Antiseptic Surgery. A Practical Treatise for the Use of Students and the General Practitioner. Illustrated with over two hundred fine Engravings. 8vo. Cloth, $5.00; sheep, $6.00.

GIBSON-RUSSELL. Physical Diagnosis: A Guide to Methods of Clinical Investigation. By G. A. Gibson, M. D., and William Russell, M. D. With 101 Illustrations. 12mo. (STUDENT'S SERIES.) Cloth, $2.50.

GOULEY (JOHN W. S.). Diseases of the Urinary Apparatus. Part I. Phlegmasic Affections. Being a Series of Twelve Lectures delivered during the autumn of 1891. With an Addendum on Retention of Urine from Prostatic Obstruction in Elderly Men. Cloth, $1.50.

GROSS (SAMUEL W.). A Practical Treatise on Tumors of the Mammary Gland. Illustrated. 8vo. Cloth, $2.50.

GRUBER (JOSEF). A Text-Book of the Diseases of the Ear. Translated from the second German edition by special permission of the author, and edited by Edward Law, M. D., and Coleman Jewell, M. D. With 165 Illustrations and 70 Colored Figures on Two Lithographic Plates. 8vo. Cloth,

GUTMANN (EDWARD). The Watering-Places and Mineral Springs of Germany, Austria, and Switzerland. Illustrated. 12mo. Cloth, $2.50.

HAMMOND (W. A.). A Treatise on Diseases of the Nervous System. With the Collaboration of Graeme M. Hammond, M. D. With One Hundred and Eighteen Illustrations. **Ninth edition**, with.corrections and additions. 8vo. Cloth, $5.00; sheep, $6.00.

AMMOND (W. A.). A Treatise on Insanity, in its Medical Relations. 8vo. Cloth, $5.00; sheep, $6.00.

HAMMOND (W. A.). Clinical Lectures on Diseases of the Nervous System. Delivered at Bellevue Hospital Medical College. Edited by T. M. B. Cross, M. D. 8vo. Cloth, $3.50.

HARVEY (A.). First Lines of Therapeutics. 12mo. Cloth, $1.50.

HIRT (LUDWIG). The Diseases of the Nervous System. A Text-Book for Physicians and Students. Translated, with permission of the Author, by August Hoch, M. D., assisted by Frank R. Smith, A. M. (Cantab.), M. D., Assistant Physicians to the Johns Hopkins Hospital. With an Introduction by William Osler, M. D., F. R. C. P., Professor of Medicine in the Johns Hopkins University, and Physician-in-Chief to the Johns Hopkins Hospital, Baltimore. 8vo, 671 pages. With 178 Illustrations. Cloth, $5.00; sheep, $6.00.

HOFFMANN-ULTZMANN. Analysis of the Urine, with Special Reference to Diseases of the Urinary Apparatus. By M. B. Hoffmann, Professor in the University of Grätz, and R. Ultzmann, Tutor in the University of Vienna. **Third edition, revised and enlarged.** 8vo. Cloth, $2.00.

HOLT (L. EMMETT). The Care and Feeding of Children. A Catechism for the Use of Mothers and Children's Nurses. 16mo. Cloth, 50 cents.

HOWE (JOSEPH W.). Emergencies, and how to treat them. Fourth edition, revised. 8vo. Cloth, $2.50.

HOWE (JOSEPH W.). The Breath, and the Diseases which give it a Fetid Odor. With Directions for Treatment. **Second edition**, revised and corrected. 12mo. Cloth, $1.00.

HUEPPE (FERDINAND). The Methods of Bacteriological Investigation. Written at the request of Dr. Robert Koch. Translated by Hermann M. Biggs, M. D. Illustrated. 8vo. Cloth, $2.50.

JACCOUD (S.). The Curability and Treatment of Pulmonary Phthisis. Translated and edited by Montagu Lubbock, M. D. 8vo. Cloth, $4.00.

JOHNSTONE (ALEX.). Botany: A Concise Manual for Students of Medicine and Science. With 164 Illustrations and a Series of Floral Diagrams. 12mo. (STUDENT'S SERIES.) Cloth, $1.75.

JONES (H. MACNAUGHTON). Practical Manual of Diseases of Women and Uterine Therapeutics. For Students and Practitioners. 188 Illustrations. 12mo. Cloth, $3.00.

JOURNAL OF CUTANEOUS AND GENITO-URINARY DISEASES. Published Monthly. Edited by John A. Fordyce, M. D. Terms, $2.50 per annum.

KEYES (E. L.). A Practical Treatise on Genito-Urinary Diseases, including Syphilis. . Being a new edition of a work with the same title by Van Buren and Keyes. Almost entirely rewritten. 8vo. With Illustrations. Cloth, $5.00; sheep, $6.00.

KEYES (E. L.). The Tonic Treatment of Syphilis, including Local Treatment of Lesions. 8vo. Cloth, $1.00.

KINGSLEY (N. W.). A Treatise on Oral Deformities as a Branch of Mechanical Surgery. With over 350.Illustrations. 8vo. Cloth, $5.00; sheep, $6.00.

LEGG (J. WICKHAM). On the Bile, Jaundice, and Bilious Diseases. With Illustrations in Chromolithography. 8vo. Cloth, $6.00; sheep, $7.00.

LITTLE (W. J.). Medical and Surgical Aspects of In-Knee (Genu-Valgum): Its Relation to Rickets, its Prevention, and its Treatment, with and without Surgical Operation. Illustrated by upward of Fifty Figures and Diagrams. 8vo. Cloth, $2.00.

LORING (EDWARD G.). A Text-Book of Ophthalmoscopy.
Part I. The Normal Eye, Determination of Refraction, and Diseases of the Media. With 131 Illustrations, and 4 Chromolithographs. 8vo. Buckram, $5.00.

Part II. Diseases of the Retina, Optic Nerve, and Choroid: their Varieties and Complications. The manuscript of this volume, which the author finished just prior to his death, has been thoroughly edited and revised by F. B. Loring, M.D., of Washington, D. C., and is now issued in the same style as the first volume. Profusely illustrated. Part II, buckram, $5.00. Two Parts, buckram, $10.00.

LUSK (WILLIAM T.). The Science and Art of Midwifery. With 246 Illustrations. **Fourth edition, revised and enlarged.** 8vo. Cloth, $5.00; sheep, $6.00.

MARCY (HENRY O.). The Anatomy and Surgical Treatment of Hernia. 4to, with about Sixty full-page Heliotype and Lithographic Reproductions from the Old Masters, and numerous Illustrations in the Text. *(Sold only by subscription.)*

MARKOE (T. M.). A Treatise on Diseases of the Bones. With Illustrations. 8vo. Cloth, $4.50.

MATHEWS (JOSEPH M.). A Treatise on Diseases of the Rectum, Anus, and Sigmoid Flexure. 8vo. With Six Chromolithographs, and Illustrations in the text. *(Sold only by subscription.)*

MILLS (WESLEY). A Text-Book of Animal Physiology, with Introductory Chapters on General Biology and a full Treatment of Reproduction for Students of Human and Comparative Medicine. 8vo. With 505 Illustrations. Cloth, $5.00; sheep, $6.00.

MILLS (WESLEY). A Text-Book of Comparative Physiology. For Students and Practitioners of Veterinary Medicine. Small 8vo. Cloth, $3.00.

MORROW (PRINCE A.). A System of Genito-Urinary Diseases, Syphilology, and Dermatology. By various Authors. In Three Volumes, beautifully illustrated. Vol. I. Genito-urinary Diseases. Vol. II. Syphilography Vol. III. Dermatology. *(Sold only by subscription.)*

THE NEW YORK MEDICAL JOURNAL (Weekly). Edited by Frank P. Foster, M.D. Terms, $5.00 per annum.
Binding Cases, cloth, 50 cents each.
"Self-Binder" (this is used for temporary binding only), 90 cents.
GENERAL INDEX, from April, 1865, to June, 1876 (23 vols.). 8vo. Cloth, 75 cts.

NIEMEYER (FELIX VON). A Text-Book of Practical Medicine, with particular reference to Physiology and Pathological Anatomy. Containing all the author's Additions and Revisions in the eighth and last German edition. Translated by George H. Humphreys, M. D., and Charles E. Hackley, M. D. 2 vols., 8vo. Cloth, $9.00; sheep, $11.00.

NIGHTINGALE'S (FLORENCE) Notes on Nursing. 12mo. Cloth, 75 cents.

OSLER (WILLIAM). Lectures on the Diagnosis of Abdominal Tumors. Small 8vo. Illustrated. Cloth, $1.50.

OSLER (WILLIAM). The Principles and Practice of Medicine. Designed for the Use of Practitioners and Students of Medicine. **Second edition, revised and enlarged.** Cloth, $5.50; sheep, $6.50; half morocco, $7.00. (*Sold only by subscription.*)

PELLEW (C. E.). A Manual of Practical Medical Chemistry. 12mo. With Illustrations. Cloth, $2.50.

PEYER (ALEXANDER). An Atlas of Clinical Microscopy. Translated and edited by Alfred C. Girard, M. D. First American, from the manuscript of the second German edition, with Additions. Ninety Plates, with 105 Illustrations, Chromolithographs. Square 8vo. Cloth, $6.00.

PIFFARD (HENRY G.). A Practical Treatise on Diseases of the Skin. By Henry G. Piffard, A. M., M. D., assisted by Robert M. Fuller, M. D. With Fifty full-page Original Plates and Thirty-three Illustrations in the Text. 4to. (*Sold only by subscription.*)

POMEROY (OREN D.). The Diagnosis and Treatment of Diseases of the Ear. With One Hundred Illustrations. **Second edition,** revised and enlarged. 8vo. Cloth, $3.00.

POORE (C. T.). Osteotomy and Osteoclasis, for the Correction of Deformities of the Lower Limbs. 50 Illustrations. 8vo. Cloth, $2.50.

QUAIN (RICHARD). A Dictionary of Medicine, including General Pathology, General Therapeutics, Hygiene, and the Diseases peculiar to Women and Children. By Various Writers. Edited by Sir Richard Quain, Bart., M. D., LL. D., etc. Assisted by Frederick Thomas Roberts, M. D., B. Sc., and J. Mitchell Bruce, M. A., M. D. With an American Appendix by Samuel Treat Armstrong, Ph. D., M. D. In two volumes. (*Sold only by subscription.*)

RANNEY (AMBROSE L.). Applied Anatomy of the Nervous System, being a Study of this Portion of the Human Body from a Standpoint of its General Interest and Practical Utility, designed for Use as a Text-Book and as a Work of Reference. **Second edition, revised and enlarged.** Profusely illustrated. 8vo. Cloth, $5.00; sheep, $6.00.

ROBINSON (A. R.). A Manual of Dermatology. Revised and corrected. 8vo. Cloth, $5.00.

ROSCOE-SCHORLEMMER. Treatise on Chemistry.

Vol. 1. Non-Metallic Elements. 8vo. Cloth, $5.00.

Vol. 2. Part I. Metals. 8vo. Cloth, $3.00.

Vol. 2. Part II. Metals. 8vo. Cloth, $3.00.

Vol. 3. Part I. The Chemistry of the Hydrocarbons and their Derivatives. 8vo. Cloth, $5.00.

Vol. 3. Part II. The Chemistry of the Hydrocarbons and their Derivatives. 8vo. Cloth, $5.00.

Vol. 3. Part III. The Chemistry of the Hydrocarbons and their Derivatives. 8vo. Cloth, $3.00.

Vol. 3. Part IV. The Chemistry of the Hydrocarbons and their Derivatives. 8vo. Cloth, $3.00.

Vol. 3. Part V. The Chemistry of the Hydrocarbons and their Derivatives. 8vo. Cloth, $3.00.

ROSENTHAL (I.). General Physiology of Muscles and Nerves. With 75 Woodcuts. 12mo. Cloth, $1.50.

SAYRE (LEWIS A.). Practical Manual of the Treatment of Club-Foot. **Fourth edition, enlarged and corrected.** 12mo. Cloth, $1.25.

SAYRE (LEWIS A.). Lectures on Orthopedic Surgery and Diseases of the Joints, delivered at Bellevue Hospital Medical College. **New edition,** illustrated with 324 Engravings on Wood. 8vo. Cloth, $5.00; sheep, $6.00.

SCHULTZE (B. S.). The Pathology and Treatment of Displacements of the Uterus. Translated from the German by Jameson J. Macan, M. A., etc.; and edited by Arthur V. Macan, M. B., etc. With one hundred and twenty Illustrations. 8vo. Cloth, $3.50.

SHIELD (A. MARMADUKE). Surgical Anatomy for Students. 12mo. (STUDENT'S SERIES.) Cloth, $1.75.

SHOEMAKER (JOHN V.). A Text-Book of Diseases of the Skin. Six Chromolithographs and numerous Engravings. Second edition, revised and enlarged. 8vo. Cloth, $5.00; sheep, $6.00.

SIMPSON (JAMES Y.). Selected Works: Anæsthesia, Diseases of Women. 3 vols.. 8vo. Per volume. Cloth, $3.00; sheep, $4.00.

SIMS (J. MARION). The Story of my Life. Edited by his Son, H. Marion-Sims, M. D. With Portrait. 12mo. Cloth, $1.50.

SKENE (ALEXANDER J. C.). A Text-Book on the Diseases of Women. Illustrated with two hundred and fifty-four Illustrations, of which one hundred and sixty-five are original, and nine chromolithographs. Second edition. 8vo. (*Sold only by subscription.*)

SKENE (ALEXANDER J. C.). Medical Gynecology. A Treatise on the Diseases of Women from the Standpoint of the Physician. 8vo. With Illustrations. Cloth, $5.00.

STEINER (JOHANN). Compendium of Children's Diseases: a Hand-Book for Practitioners and Students. Translated from the second German edition, by Lawson Tait. 8vo. Cloth, $3.50; sheep, $4.50.

STEVENS (GEORGE T.) Functional Nervous Diseases: their Causes and their Treatment. Memoir for the Concourse of 1881-1883, Académie Royal de Médecine de Belgique. With a Supplement, on the Anomalies of Refraction and Accommodation of the Eye, and of the Ocular Muscles. Small 8vo. With six Photographic Plates and twelve Illustrations. Cloth, $2.50.

STONE (R. FRENCH). Elements of Modern Medicine, including Principles of Pathology and of Therapeutics, with many Useful Memoranda and Valuable Tables of Reference. Accompanied by Pocket Fever Charts. Designed for the Use of Students and Practitioners of Medicine. In wallet-book form, with pockets on each cover for Memoranda, Temperature Charts, etc. Roan, tuck, $2.50.

STRECKER (ADOLPH). Short Text-Book ot Organic Chemistry. By Dr. Johannes Wislicenus. Translated and edited, with Extensive Additions, by W. H. Hodgkinson and A. J. Greenaway. 8vo. Cloth, $5.00.

STRÜMPELL (ADOLPH). A Text-Book of Medicine, for Students and Practitioners. Translated, by permission, from the sixth German edition by Herman F. Vickery, A. B., M. D., Instructor in Clinical Medicine, Harvard Medical School, etc., and Philip Coombs Knapp, Physician to Outpatients with Diseases of the Nervous System, Boston City Hospital, etc. With Editorial Notes by Frederick C. Shattuck, A. M., M. D., Jackson Professor of Clinical Medicine, Harvard Medical School, etc. Second American edition. With 111 Illustrations. 8vo. 981 pages. Cloth, $6.00; sheep, $7.00.

THOMAS (T. GAILLARD). Abortion and its Treatment, from the Standpoint of Practical Experience. A Special Course of Lectures delivered before the College of Physicians and Surgeons, New York, Session of 1889-'90. From Notes by P. Brynberg Porter, M. D. Revised by the Author. 12mo. Cloth, $1.00.

THOMPSON (W. GILMAN). Practical Dietetics, with Special Reference to Diet in Disease. 8vo. Cloth, $5.00.

THOMSON (J. ARTHUR). Outlines of Zoölogy. With thirty-two full-page Illustrations. 12mo. (STUDENTS' SERIES.) Cloth, $3.00.

TILLMANNS (HERMANN). The Principles of Surgery and Surgical Pathology. Translated by John Rogers, M. D., and Benjamin Tilton, M. D., New York. 8vo. With 441 Illustrations. Cloth, $5.00; sheep, $6.00.

ULTZMANN (ROBERT). Pyuria, or Pus in the Urine, and its Treatment. Translated by permission, by Dr. Walter B. Platt. 12mo. Cloth, $1.00.

VAN BUREN (W. H.). Lectures upon Diseases of the Rectum, and the Surgery of the Lower Bowel, delivered at Bellevue Hospital Medical College. **Second edition, revised and enlarged.** 8vo. Cloth, $3.00; sheep, $4.00.

VAN BUREN (W. H.). Lectures on the Principles and Practice of Surgery. Delivered at Bellevue Hospital Medical College. Edited by Lewis A. Stimson, M. D. 8vo. Cloth, $4.00; sheep, $5.00.

VOGEL (A.). A Practical Treatise on the Diseases of Children. Translated and edited by H. Raphael, M. D. **Third American from the eighth German edition, revised and enlarged.** Illustrated by six Lithographic Plates. 8vo. Cloth, $4.50; sheep, $5.50.

VON ZEISSL (HERMANN). Outlines of the Pathology and Treatment of Syphilis and Allied Venereal Diseases. **Second edition,** revised by Maximilian von Zeissl. Authorized edition. Translated, with Notes, by H. Raphael, M. D. 8vo. Cloth, $4.00; sheep, $5.00.

WAGNER (RUDOLF). Hand-Book of Chemical Technology. Translated and edited from the eighth German edition, with extensive Additions. by William Crookes. With 336 Illustrations. 8vo. Cloth, $5.00.

WALTON (GEORGE E.). Mineral Springs of the United States and Canadas. Containing the latest Analyses, with full Description of Localities, Routes, etc. **Second edition, revised and enlarged.** 12mo. Cloth, $2.00.

WEBBER (S. G.). A Treatise on Nervous Diseases: Their Symptoms and Treatment. A Text-Book for Students and Practitioners. 8vo. Cloth, $3.00.

WEEKS-SHAW (CLARA S.). A Text-Book of Nursing. For the Use of Training-Schools, Families, and Private Students. Second edition, revised and enlarged. 12mo. With Illustrations, Questions for Review and Examination, and Vocabulary of Medical Terms. 12mo. Cloth, $1.75.

WELLS (T. SPENCER). Diseases of the Ovaries. 8vo. Cloth, $4.50.

WORCESTER (A.). Monthly Nursing. **Second edition, revised.** Cloth, $1.25.

WYETH (JOHN A.). A Text-Book on Surgery: General, Operative, and Mechanical. Profusely illustrated. **Second edition, [revised and enlarged.** 8vo. (*Sold only by subscription.*)

www.ingramcontent.com/pod-product-compliance
Lightning Source LLC
Chambersburg PA
CBHW021733220426
43662CB00008B/829